FUEL CELLS FOR BUILDING APPLICATIONS

This publication was prepared under ASHRAE Research Project 1058-RP in cooperation with TC 9.5, Cogeneration Systems.

About the Author

Michael W. Ellis, Ph.D., P.E., *Member ASHRAE*, is an assistant professor in the Department of Mechanical Engineering at Virginia Polytechnic Institute and State University. His research interests include modeling and optimization of fuel cell systems and energy conservation in commercial and industrial facilities. Dr. Ellis joined the faculty at Virginia Tech after earning his doctorate in mechanical engineering from the Georgia Institute of Technology in 1996. Prior to graduate school he worked for six years as a mechanical design engineer and project manager for an engineering/architecture firm.

FUEL CELLS FOR BUILDING APPLICATIONS

Michael W. Ellis

American Society of Heating, Refrigerating
and Air-Conditioning Engineers, Inc.

ISBN 1-931862-03-6

1791 Tullie Circle, N.E.
Atlanta, GA 30329
www.ashrae.org

Printed in the United States of America

Cover photograph provided by UTC Fuel Cells.

ASHRAE Staff

Special Publications

Mildred Geshwiler
Editor

Erin Howard
Assistant Editor

Christina Johnson
Editorial Assistant

Michshell Phillips
Secretary

Publishing Services

Barry Kurian
Manager

Jayne Jackson
Production Assistant

Publisher

W. Stephen Comstock

TABLE OF CONTENTS

PREFACE

Fuel cell systems are a promising new type of on-site power generation technology that offers modularity, efficient operation across a wide range of load conditions, minimal environmental impact, and opportunities for integration into cogeneration systems. Cogeneration systems simultaneously provide power and heat and offer a number of advantages including economic benefits, better utilization of energy resources, and reduced environmental impact. The advent of fuel cell cogeneration systems promises to make these advantages available in applications that were previously not well suited for cogeneration.

There are four types of fuel cell technology that are applicable for building systems. Proton exchange membrane fuel cells (PEMFC) operate at low temperatures and are applicable in sizes as small as a few kWe. Phosphoric acid fuel cells (PAFC) have been demonstrated in several hundred building applications, typically at sizes of approximately 200 kWe. Molten carbonate fuel cells (MCFC) operate at higher temperatures and offer more flexibility in fuel processor design and integration. Solid oxide fuel cells (SOFC) operate at temperatures as high as 1000°C (1800°F) and offer flexible fuel processor design and integration as well as opportunities for additional electricity production using the high-temperature exhaust stream.

Each of the four technologies is currently the subject of extensive research and development efforts. Research is underway to improve the performance, extend the operating life, reduce the initial cost, and improve the tolerance to fuel stream impurities. These goals are being accomplished through improvements in materials, optimization of operating conditions, and advances in manufacturing methods. A variety of demonstration projects are underway for each technology. These projects provide valuable data on the actual field performance of these systems.

Fuel cell systems include not only the fuel cell itself but also the fuel processor, power conditioner, thermal management, and water management subsystems. These different subsystems are integrated to form fuel cell systems that yield electrical conversion efficiencies that can exceed 45 percent. Furthermore, fuel cell systems can maintain high efficiency at loads as low as 50 percent of full load. Thus, fuel cell systems exhibit performance characteristics that are comparable to the most efficient stationary power systems and provide this performance over a broad range of loads and in relatively small sizes.

Emissions from the fuel cell system occur primarily from the fuel processor and are usually far below regulatory limits. With low emissions, minimal noise, and a relatively unobtrusive appearance, fuel cell systems can be sited in a variety of locations near the point of power use. The life expectancy of a fuel cell system is projected to be twenty years, with replacement of the fuel cell stack required at roughly five-year intervals. Other than stack replacement, fuel cell system maintenance consists of the routine maintenance procedures that are required for the fans, pumps, and controls that comprise the support systems. Estimates of annual maintenance costs, including a sinking fund for stack replacement, range from less that $0.01/kWh to as high a $0.03/kWh.

Application of fuel cell cogeneration systems in buildings requires selection of an operating strategy, coordination with the local utility systems, and evaluation of the economic value of the system. A simple calculation of the cost of electricity provided by the fuel cell system can provide a preliminary indication of the economic potential. A more detailed hour by hour model can provide a more accurate estimate of the savings associated with a fuel cell cogeneration system. Simple cost of electricity calculations indicate that fuel cell systems can be competitive in residential and commercial applications provided initial costs can be reduced to approximately $1,500/kWe. Current projections indicate that fuel cell systems can be provided at costs below $2,000/kWe with some projections indicating costs as low as $500/kWe. Once the economic criteria are satisfied, implementation of fuel cell cogeneration systems will require meeting the requirements of building codes and standards. Currently, a variety of codes and standards are under development, including codes governing installation in buildings, standards for fuel cell system construction and performance testing, and standards for interconnection with the local utility.

Considerable progress is being made in the development and commercialization of fuel cell cogeneration systems for buildings. Progress is being made in the development of fuel cell technology, system integration, manufacturing methods, and in the development of a regulatory infrastructure to facilitate the implementation of fuel cell technology. Building designers are a critical link in the dissemination of this new technology. By understanding fuel cells and cogeneration systems, designers can identify appropriate applications that will yield economic and environmental benefits for their clients and for society.

Chapter 1

INTRODUCTION

This technology assessment guide is intended to introduce the reader to the basic principles of fuel cell technology, to assess the current status of fuel cell technology for building applications, and to provide an understanding of fuel cell systems and their application in building cogeneration systems. Cogeneration systems simultaneously provide both electrical and thermal energy. Cogeneration systems based on conventional generating technologies, such as reciprocating engines, gas turbines, and steam turbines, have been employed in industrial facilities, district heating systems, and, to a limited extent, in large commercial buildings. From an energy use and environmental impact point of view, cogeneration systems are very attractive. In a traditional electric power plant, only about 30% to 40% of the fuel energy is converted to electric power, and the remainder is discharged to the atmosphere or a nearby river. With cogeneration, this thermal energy can be recovered and used for space heating, water heating, manufacturing processes, etc. Typically, an additional 30% to 40% of the fuel energy can be utilized. This means that 30% to 40% less primary energy is used, and emissions of CO_2 and other pollutants can be reduced.

The widespread application of cogeneration systems in buildings has been limited by several factors:

- Conventional generating technology tends to be more economical in large sizes.
- Conventional generating technology is most efficient in large sizes and when operating near full load.
- Conventional generating technology tends to require a larger, more skilled maintenance staff than is available in many building applications.
- Noise and environmental emissions restrict the potential sites for conventional generating technology.

Fuel cell-based cogeneration systems overcome many of these limitations and greatly expand the possible applications for cogeneration.

Chapter 1 of this guide provides a discussion of fuel cell fundamentals and introduces the four major types of fuel cells that are most applicable in buildings. Chapter 2 provides a detailed discussion of the current status of each of the fuel cell technologies with respect to design, manufacturing, and commercialization. Chapter 3 discusses the integration of the fuel cell into a fuel cell system that typically includes a fuel processor, power conditioner, and thermal, water, and air management subsystems. Chapter 3 also reviews the system characteristics that are most significant to building applications. Chapter 4 discusses the application of fuel cell cogeneration systems in buildings and highlights the economic, regulatory, and application issues that must be considered when evaluating these systems. Chapter 5 summarizes the material presented in the guide and discusses the future of fuel cell cogeneration systems in buildings.

FUEL CELL FUNDAMENTALS

Fuel Cell Description

Fuel cells are electrochemical devices that convert chemical energy to electrical energy and heat. In a fuel cell the conversion process from chemical energy to electricity is direct. In contrast, conventional energy conversion processes first transform chemical energy to heat through combustion and then convert heat to electricity through some type of power cycle (e.g., steam power plant, gas turbine, or internal combustion engine) coupled with a generator. The direct nature of the fuel cell process yields relatively high efficiencies and, in many cases, simplified power systems when compared to conventional energy conversion systems.

A schematic of a fuel cell is illustrated in Figure 1-1. The fuel cell consists of the following five major elements:

1. Fuel flow channel
2. Anode
3. Electrolyte
4. Cathode
5. Oxidant flow channel

Fuel enters the cell through the fuel flow channel and then travels into the anode where it is oxidized, yielding electrons that travel through the external electrical load. Oxidant enters the cell through the oxidant flow

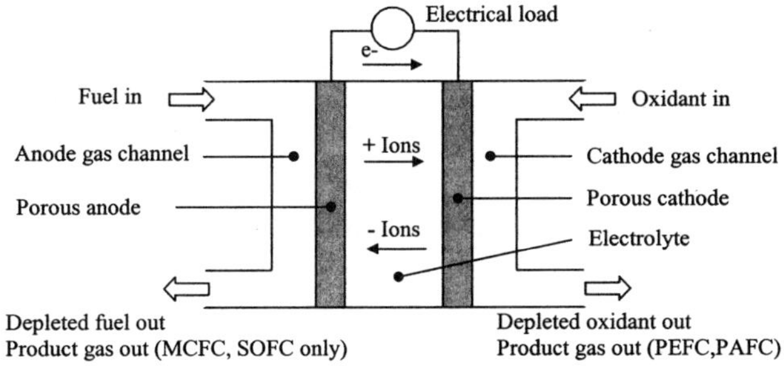

Fuel cell type	+ Ion	- Ion
PEFC	H^+	none
PAFC	H^+	none
MCFC	none	CO_3^{2-}
SOFC	none	O^{2-}

Figure 1-1 Fuel cell schematic.

channel and then travels into the cathode where it is reduced as it reacts with electrons from the external circuit. Ions are exchanged between the anode and the cathode to balance the reaction. The details of the half reactions and the composition of the ions travelling within the electrolyte depend on the type of fuel cell, but the net fuel cell reaction is:

$$H_2 + 1/2\ O_2 \rightarrow H_2O \tag{1-1}$$

Fuel cells are most commonly classified by the type of electrolyte and include

- proton exchange membrane fuel cells (PEMFC),[1]
- alkaline fuel cells (AFC),
- phosphoric acid fuel cell (PAFC),
- molten carbonate fuel cells (MCFC),
- solid oxide fuel cells (SOFC), and
- direct methanol fuel cells (DMFC).

1. Proton exchange membrane fuel cells (PEMFC) are also referred to as polymer electrolyte fuel cells (PEFC).

The AFC is of historical interest as it was one of the first types to see practical use, serving as the primary power source on the Apollo space missions and on the space shuttle. However, this type of fuel cell is being phased out in favor of the PEMFC for space applications and is of limited value in terrestrial applications (where the oxidant is air) because of the sensitivity of the AFC electrolyte to CO_2. In transportation and portable power applications, the portability of liquid methanol makes the direct methanol fuel cell attractive. In stationary applications, methanol is not as attractive as more readily available fuels, such as natural gas and LP gas, which are used in the other four fuel cell types. This book will focus on the remaining four cell types—PEMFC, PAFC, MCFC, and SOFC—that are being actively developed for stationary applications.

Thermodynamics of Fuel Cell Operation

The reversible work associated with a chemical reaction involving N moles of fuel is given by the change in the Gibbs function,

$$W = -N\Delta\overline{G}, \tag{1-2}$$

where $\Delta\overline{G}$ is the change in Gibbs function per mole of fuel.

For a reaction among ideal gases given by

$$\alpha A + \beta B \rightarrow \chi C + \delta D, \tag{1-3}$$

the Gibbs function change per mole of fuel (i.e., reactant A) is given by

$$\Delta\overline{G} = \Delta\overline{G}°(T) + RT\ln\frac{p_C^{\chi}p_D^{\delta}}{p_A^{\alpha}p_B^{\beta}} \tag{1-4}$$

where $\Delta\overline{G}°(T)$ is the Gibbs function change per mole of fuel at standard pressure and the indicated temperature, and p is the partial pressure of each constituent in atmospheres.

The work associated with an electrochemical reaction can also be determined by the amount of charge that is moved through the electrical potential, E, associated with the reaction

$$W = nNFE \tag{1-5}$$

where n is the number of electrons involved per mole of fuel and F is Farraday's constant. Combining Equations 1-2 and 1-5 yields

$$E = \frac{-\Delta\overline{G}}{nF}. \tag{1-6}$$

Incorporating Equation 1-4 for a reaction among ideal gases gives the open circuit potential,

$$E = E°(T) - \frac{RT}{nF}\ln\frac{p_C^{\chi}p_D^{\delta}}{p_A^{\alpha}p_B^{\beta}} \tag{1-7}$$

where E°(T) is the open circuit potential at standard pressure and the indicated temperature, defined by

$$E°(T) = \frac{-\Delta G°(T)}{nF}. \tag{1-8}$$

The open circuit potential given by Equation 1-7 represents the maximum voltage between the anode and cathode for the given cell reaction at the indicated temperature and reactant pressure. For substances other than ideal gases, Equation 1-7 can be generalized as

$$E = E°(T) - \frac{RT}{nF}\ln\frac{a_C^{\chi}a_D^{\delta}}{a_A^{\alpha}a_B^{\beta}} \tag{1-9}$$

where the a_i represent the activities of the products and reactants. Equation 1-9 is the general form of the Nernst equation [1].

Equation 1-6 can be used to derive an expression indicating the variation of open circuit potential with temperature and pressure. The Gibbs function is defined as

$$G = H - TS. \tag{1-10}$$

Taking the derivative of this expression using the chain rule yields

$$dG = dH - TdS - SdT. \tag{1-11}$$

Substituting the second Gibbs equation,

$$TdS = dH - Vdp, \tag{1-12}$$

into Equation 1-11 yields

$$dG = -SdT + Vdp \tag{1-13}$$

Equation 1-13 applies to both reactants and products:

$$dG_P = -S_PdT_P + V_Pdp_P \tag{1-14a}$$

$$dG_R = -S_R dT_R + V_R dp_R \tag{1-14b}$$

If the products and reactants are at the same temperature and pressure, Equations 1-14a and 1-14b can be combined with Equation 1-6 to yield

$$\left.\frac{\partial E}{\partial T}\right)_{\substack{constant\\pressure}} = \frac{\Delta S}{nF} \tag{1-15a}$$

and

$$\left.\frac{\partial E}{\partial p}\right)_{\substack{constant\\temperature}} = -\frac{\Delta V}{nF} \tag{1-15b}$$

where ΔS and ΔV are the changes in entropy and volume, respectively, that are associated with the chemical reaction.

For example, for the reaction of hydrogen and oxygen to produce water, Equation 1-1, the change in Gibbs function, entropy, and volume at standard conditions of 25°C (77°F) and 1 atm are given by:

$$\Delta G = -229{,}000 \text{ J/gmol} \tag{1-16a}$$

$$\Delta S = -44 \text{ J/gmol-K} \tag{1-16b}$$

$$\Delta V = -0.0123 \text{ m}^3/\text{ g mol-K} \tag{1-16c}$$

when the product water is in the gas phase.

Thus, for a fuel cell operating on hydrogen and oxygen with the product water in the gas phase, Equation 1-7 yields an open circuit voltage of

$$E = \frac{229{,}000 \dfrac{\text{J}}{\text{gmol} - \text{H}_2}}{2\dfrac{\text{gmol} - \text{e}^-}{\text{gmol} - \text{H}_2} \times 96{,}487 \dfrac{\text{Coul}}{\text{gmol} - \text{e}^-}}$$

$$-\frac{8.314 \dfrac{\text{J}}{\text{gmol} - \text{H}_2 \times \text{K}} \times 298\text{K}}{2\dfrac{\text{gmol} - \text{e}^-}{\text{gmol} - \text{H}_2} \times 96{,}487 \dfrac{\text{Coul}}{\text{gmol} - \text{e}^-}} \times \ln \frac{(1\text{atm})^1}{(1\text{atm})^1 \times (1\text{atm})^{0.5}} \tag{1-17}$$

$$E = 1.18 \frac{\text{J}}{\text{Coul}} - 0.0128 \times \ln(1) \frac{\text{J}}{\text{Coul}} = 1.18 \text{ V.}$$

This is the voltage that would be measured across the fuel cell when there is no current flow.

Since ΔS is negative, Equation 1-15a indicates that the open circuit voltage decreases with increasing temperature. Since the change in volume due to the reaction is negative, Equation 1-15b indicates that the open circuit voltage increases with increasing pressure. Figure 1-2 illustrates the variation of open circuit voltage with temperature. Neglecting other factors, this would predict a higher efficiency for hydrogen-fueled cells operating at a lower temperature. However, when current flows through the cell, other effects, called polarization losses, reduce the cell voltage below the open circuit value. These polarization losses tend to be reduced at elevated temperatures, thus counteracting the trend illustrated in Figure 1-2.

Polarization Losses

To produce work, current must flow between the anode and cathode. The flow of current leads to a reduction in cell voltage that is referred to as the *polarization loss* or *overvoltage*. The polarization loss is generally divided into three components—activation polarization, ohmic polarization, and concentration polarization.

Activiation polarization refers to the voltage drop required to move the electrochemical reaction forward at a finite rate. At both the anode and cathode, reactants diffuse to reaction sites where the reduction and oxidation reactions take place. At the reaction sites, an activation potential (or voltage difference) must be overcome before the reaction can proceed. Catalysts are used in the low-temperature fuel cells to lower the activation potential and allow the reaction to proceed. At higher

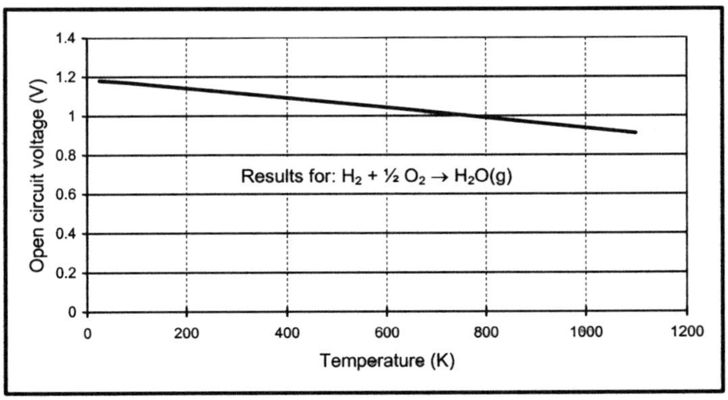

Figure 1-2 Variation of open circuit potential with temperature.

temperatures, the activation potential is reduced and there is less need for catalysts.

When the current density is very low, the required reactant transport rates are also low and, thus, the concentration gradients across the fuel cell are small. In this case, the reaction rate is controlled by the activation overpotential and the concentrations of reactants at the reaction sites approach the inlet concentrations. For this case, the activation polarization is commonly expressed as a function of the current density using an empirical equation called the Tafel equation,

$$V_{act} = a\log(i/i_0)\,, \tag{1-18}$$

where the Tafel slope, a, and the exchange current density, i_0, are experimentally determined constants [1].

As this equation indicates, the activation polarization increases logarithmically with current density. Thus the activation polarization changes rapidly at low current density and then changes more slowly at higher current densities. The Tafel slope and the exchange current density determine the activation polarization for a particular current density. Catalysts reduce the value of the Tafel slope and increase the exchange current density, thus decreasing the activation polarization for a given current density.

Ohmic polarization refers to the voltage drop that is attributable to the resistance to the flow of electrons and ions through the cell. Ohmic losses are roughly a linear function of the cell current:

$$V_{ohm} = iR \tag{1-19}$$

Concentration polarization refers to the voltage drop that is attributable to reactant flow restrictions at the electrodes. As the current density increases, the reactant flow rates required to sustain the reaction also increase. Eventually, the rate of transport through the cell cannot keep up with the rate at which the reactants are consumed. The reactant concentration at each reaction site begins to drop and thus the voltage drops.

The net cell voltage is the open circuit voltage less the various polarization losses:

$$V_{cell} = V_{oc} - V_{act} - V_{ohm} - V_{conc} \tag{1-20}$$

The power supplied by the fuel cell, $\dot{W}$, is given by the product of the current and the cell voltage:

$$\dot{W} = V_{cell} I_{cell} \tag{1-21}$$

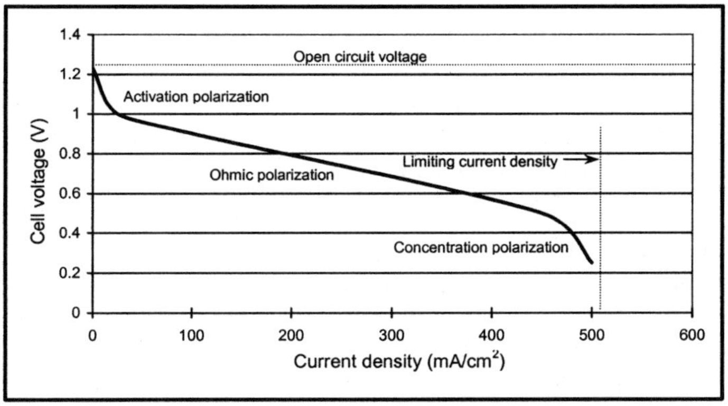

Figure 1-3 Typical fuel cell polarization curve.

Thus, the polarization losses, by reducing the cell voltage, directly reduce the electrical power available from the cell. Due to these losses, energy that would be available as electricity must instead be transferred from the cell as heat.

A polarization curve relates the cell voltage to the cell current. A typical polarization curve is presented in Figure 1-3. As the figure indicates, the open circuit voltage is the cell voltage when the current flow is zero. As current begins to flow, the voltage drops due to the activation polarization. Due to the logarithmic behavior, the activation polarization increases rapidly at first and then more slowly as the current density increases. At some point, increases in the ohmic polarization become more significant than those associated with the activation polarization. The polarization curve then takes on the linear shape associated with the ohmic losses. Finally, at high current densities, restrictions to the flow of reactants to the reaction sites lead to a limiting current density. While the values associated with the curve may change, all fuel cell technologies exhibit this general behavior. In addition to varying with the type of fuel cell, the polarization curve is affected by the fuel cell operating conditions. For example, changes in temperature and pressure can cause the curve to shift up or down and can change the value of the limiting current.

Fuel Cell Performance Measures

Fuel cell performance can be measured in terms of voltage (V), current density (mA/cm^2), power density (W/cm^2), and efficiency. The cell voltage and current density are related by a polarization curve such as the one in Figure 1-3. The power density is the product of the cell voltage and current density. At low current densities, the power density increases with increasing current. However, as the limiting current density is approached, the decreasing cell voltage becomes more significant than the increasing cell current, and the power begins to decrease. Thus, the cell can be characterized by a maximum power density.

There are a number of ways to describe the fuel cell efficiency. The fuel cell transforms chemical energy into electrical power and heat. The reversible, or *ideal efficiency,* is the maximum electrical power that could be obtained from the chemical reaction occurring in the cell divided by the rate at which chemical energy is transformed by the complete reaction of the fuel entering the cell. The ideal efficiency is given by

$$\eta_{ideal} \equiv \frac{\dot{W}_{ideal}}{N \cdot \overline{HV}_a} = \frac{V_{oc} nNF}{N \cdot \overline{HV}_a} = \frac{V_{oc} nF}{\overline{HV}_a} \tag{1-22}$$

where $\dot{W}_{ideal}$ is the ideal power, N is the molar rate of reaction of the fuel (mol/s), n is the number of electrons involved in the electrochemical reaction ($n = 2$ for H_2), F is Faraday's constant, and $\overline{HV}_a$ is either the higher or lower heating value of the fuel supplied to the anode on a molar basis. If the heating value is expressed on a mass basis, Equation 1-22 becomes

$$\eta_{ideal} = \frac{V_{oc} nF}{M_a HV_a} \tag{1-23}$$

where M_a is the molar mass of the fuel supplied to the anode.

The heating value, HV, is the amount of energy released by the complete reaction of the fuel and depends on whether the product water leaving the fuel cell is in the liquid or vapor phase. If water is produced in the liquid phase, as it is in PEMFCs, more energy is available from the reaction and the heating value is called the higher heating value (HHV). If water is produced in the vapor phase, as it is for most other fuel cells, less energy is available from the reaction and the heating value is called the lower heating value (LHV).

The ideal efficiency does not reflect the losses associated with the activation, ohmic, and concentration polarizations. These effects reduce the cell voltage below the open circuit value. A second measure of efficiency that captures these effects is the *voltage efficiency,* given by

$$\eta_v = \frac{V_{cell}}{V_{oc}}. \tag{1-24}$$

Referring to the polarization curve, it is obvious that the voltage efficiency decreases as the current density increases.

Just as the actual cell voltage is less than the ideal, the actual current is also less than the ideal current. In the ideal case, every H_2 molecule would react to yield two electrons to contribute to the current flow. In fact, some H_2 crosses through the electrolyte without reacting and some current may be shunted through the electrolyte. Thus, the actual current is less than that corresponding to the flow of H_2. The current efficiency or Farradic efficiency is given by

$$\eta_I = \frac{I_{cell}}{nNF}. \tag{1-25}$$

The concentration of reactants decreases through the cell as the reactants are consumed, while inert gases remain unchanged and product gases increase. Thus, in theory, the concentration of reactants leaving the cell could approach zero. But at very low reactant concentrations, the cell reaction cannot proceed without large polarization losses. Thus, as a practical matter, the reactant streams leaving the cell must contain a significant concentration of reactants. The fuel efficiency, or utilization, refers to the amount of fuel that is actually removed from the anode stream due to chemical reaction divided by the amount of fuel entering the anode:

$$\eta_U = \frac{N}{N_a} \tag{1-26}$$

where N_a is the molar flow rate of fuel into the anode.

The energy produced by the fuel cell is not the same as the energy actually available from the fuel cell system. Some of the fuel cell electricity is dissipated as heat from the power conditioner and some goes to operate auxiliary system equipment such as pumps and fans, thus reducing the net electricity available from the system. On the other hand, in some fuel cell systems, energy from the exhaust stream can be recovered and used to produce additional power so that the system power exceeds the power produced by the cell alone. These factors are reflected in the system efficiency, which is given by

$$\eta_s \equiv \frac{\dot{W}_S}{\dot{W}} = \frac{\dot{W}_S}{V_{cell}I_{cell}} \tag{1-27}$$

where $\dot{W}_s$ is the net electrical power produced by the overall system. The system efficiency may be greater or less than one.

If the fuel cell system includes a fuel processor, the heating value of the fuel entering the fuel processor will be different than the heating value of the fuel entering the fuel cell anode. The fuel processor efficiency is defined as [2]

$$\eta_{FP} \equiv \frac{HV_a}{HV_f} \tag{1-28}$$

where HV_f is the heating value of the fuel entering the fuel processor. Since heat is added to the fuel stream in the fuel processor, the heating value of the gas leaving the processor may actually be higher than the heating value of the fuel entering the fuel processor. Thus the fuel processor efficiency may be greater than or less than one.

The *overall electrical conversion efficiency* is the electrical energy provided by the fuel cell system divided by the energy released by the complete reaction of all of the fuel entering the cell. The overall electrical conversion efficiency is the most comprehensive measure of fuel cell system performance and includes the effects of the other six efficiency measures:

$$\eta_E \equiv \frac{\dot{W}_S}{\dot{m}_f \cdot HV_f} = \frac{V_{oc}nF}{M_a HV_a} \times \frac{V_{cell}}{V_{oc}} \times \frac{I_{cell}}{nNF} \times \frac{N}{N_a} \times \frac{\dot{W}_S}{V_{cell}I_{cell}} \times \frac{HV_a}{HV_f} \times \frac{N_a M_a}{\dot{m}_f} \tag{1-29}$$

The final term in Equation 1-29 represents the ratio of the mass flow rate of gas entering the anode to the mass flow rate of fuel entering the fuel processor. This term may be greater than one in fuel processors where air is introduced into the fuel stream to oxidize fuel or CO, or it may be less than one in fuel processors where water is condensed from the fuel stream. Designating this mass ratio term as ε_{FP} and combining Equations 1-22 and 1-29 yields

$$\eta_E \equiv \eta_{ideal}\eta_V\eta_I\eta_U\eta_S\eta_{FP}\varepsilon_{FP}. \tag{1-30}$$

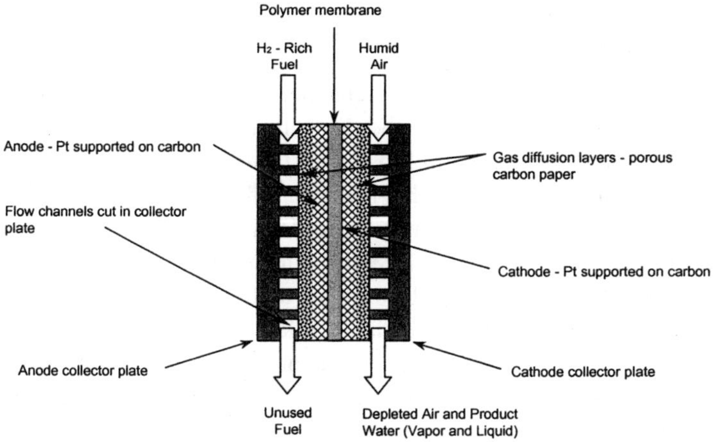

Figure 1-4 PEMFC structure.

PROTON EXCHANGE MEMBRANE FUEL CELL (PEMFC)

The proton exchange membrane fuel cell (PEMFC) is notable for a high current density and a low operating temperature. These features have made this fuel cell technology particularly attractive in transportation applications where compact size and rapid start-up are important characteristics. The possibility that this technology may achieve costs that are competitive with internal combustion engines has attracted the attention of developers interested in small-scale stationary applications.

The cell reactions for a PEMFC are given by:

$$\text{Anode:} \quad H_2 \rightarrow 2H^+ + 2e^- \tag{1-31}$$

$$\text{Cathode:} \quad 2H^+ + 2e^- + \frac{1}{2}O_2 \rightarrow H_2O \tag{1-32}$$

$$\text{Net:} \quad H_2 + \frac{1}{2}O_2 \rightarrow H_2O \tag{1-33}$$

The cell typically operates at temperatures of 50°C to 90°C (120° to 190°F). At these low temperatures, both the anode and cathode reactions require noble metal catalysts to facilitate the reaction. The cell structure for a typical PEMFC is illustrated in Figure 1-4.

The PEMFC stack uses hydrogen as its fuel. Hydrocarbon fuels such as methane and propane and others can be reformed to a hydrogen-rich gas stream that can then be used to fuel the PEMFC. Gases other than hydrogen in the fuel stream can degrade system performance. In particular, carbon monoxide and compounds of sulfur can poison the catalysts in the fuel cell stack.

The fuel gas flows through channels cut in the anode collector plate. Hydrogen diffuses from the flow channels through a porous carbon paper diffusion layer and to the surface of the anode. The anode consists of platinum catalyst that is supported on carbon black, which is bonded to the surface of the membrane. Hydrogen reacts at the catalyst sites, releasing protons that flow through the polymer membrane and electrons that travel to the collector plate and then through the electrical circuit. The polymer membrane has a relatively high protonic conductivity when saturated with water. As the protons flow through the membrane, they take water molecules with them by electro-osmotic drag.

The oxidant, typically air, flows through channels cut in the cathode collector plate. Oxygen diffuses from the flow channels through a carbon paper diffusion layer and to the surface of the cathode. The cathode construction is similar to the anode construction except that the platinum loading is typically higher. At the cathode catalyst sites, oxygen reacts with electrons from the cathode collector plate and hydrogen protons passing through the membrane producing water in liquid form. This water evaporates into the cathode gas stream, drains from the cell, or is transported to the anode by diffusion or pressure-driven flow.

The protonic conductivity of the polymer membrane is strongly dependent on the moisture content of the membrane. As the moisture content decreases, the resistivity increases. Consequently, the ohmic polarization increases and more heat must be removed from the cell. This can lead to poor performance and local regions of high temperature. Thus, control of membrane moisture content is a major issue with PEMFCs. The membrane moisture content is related to the relative humidity of the reactant gas streams. To avoid drying out the membrane, the inlet gas streams are typically heated to the operating temperature of the cell and humidified to 80% to 90% relative humidity.

Individual cells are assembled into a stack as illustrated in Figure 1-5. Bipolar plates are located between cells. The bipolar plate serves as the anode collector plate for one cell and the cathode collector plate for the adjacent cell, effectively connecting the cells in series. Cooling channels are located periodically within the stack. A cooling fluid—typically air or water—is circulated through the cooling channels to remove heat from the stack. Stack sizes range from a few watts to 250 kWe. Larger stacks typi-

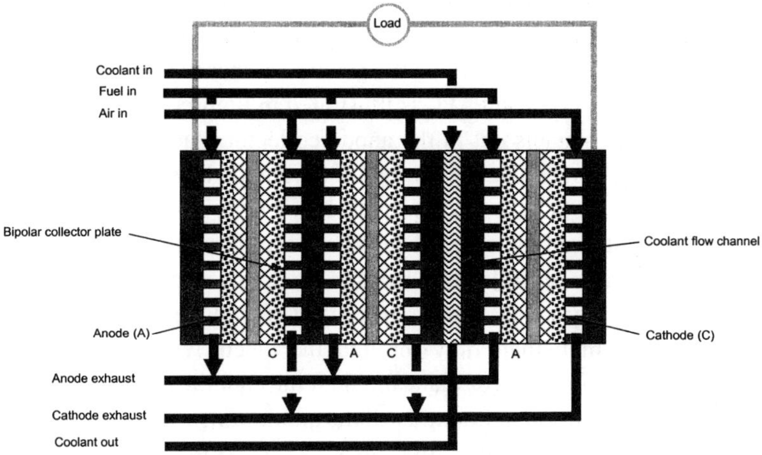

Figure 1-5 PEMFC stack configuration.

cally have a cross-sectional area that is on the order of 600 cm^2 and may contain over 100 cells.

Representative performance measures for a PEMFC stack operating with H_2 fuel are:

Cell voltage:	0.6 to 0.8 V/cell
Current density:	500 to 1,000 mA/cm^2
Power density:	0.3 to 0.8 W/cm^2

PHOSPHORIC ACID FUEL CELLS (PAFC)

Of all the fuel cell technologies, the phosphoric acid fuel cell is at the most advanced state of development. Fuel cells based on PAFC technology have been installed at roughly 300 locations worldwide. The most common application for PAFC technology is providing supplementary power for building applications. PAFCs providing 200 kWe of electric power, and up to 200 kW of thermal energy, have been installed in hotels, hospitals, office buildings, and military facilities.

The cell reactions for a PAFC are the same as those for a PEMFC:

$$\text{Anode:} \quad H_2 \rightarrow 2H^+ + 2e^- \tag{1-34}$$

$$\text{Cathode:} \quad 2H^+ + 2e^- + \frac{1}{2}O_2 \rightarrow H_2O \tag{1-35}$$

$$\text{Net:} \quad H_2 + \frac{1}{2}O_2 \rightarrow H_2O \tag{1-36}$$

The PAFC usually operates at a temperature of 200°C (390°F) and at pressures ranging from 1 to 8 atm. As with the PEMFC, noble metal catalysts are required to facilitate the reaction at these temperatures. The cell structure for a typical PAFC is illustrated in Figure 1-6. The fuel gas flows through channels cut in the anode collector plate. Hydrogen diffuses from the flow channels through a porous carbon substrate. Platinum catalyst is disbursed on carbon black in a layer located between the porous substrate and the electrolyte. The electrolyte is pure phosphoric acid supported in a PTFE-bonded silicon-carbide matrix. Hydrogen reacts at the catalyst sites that are in contact with the electrolyte releasing protons, which then flow through the electrolyte and electrons, which travel to the collector plate and then through the electrical circuit. The cathode construction is similar to the anode construction except that the platinum loading is typically higher. At the cathode catalyst sites, oxygen reacts with electrons from the cathode collector plate and hydrogen protons passing through the electrolyte to produce water in vapor form. This water diffuses into the flow channels and leaves with the depleted oxidant.

Individual cells are assembled into a stack that has a configuration similar to that of the PEMFC. Cooling channels are located periodically within the stack. Water, a cooling fluid, is typically circulated through the cooling channels to remove heat from the stack. Common stack

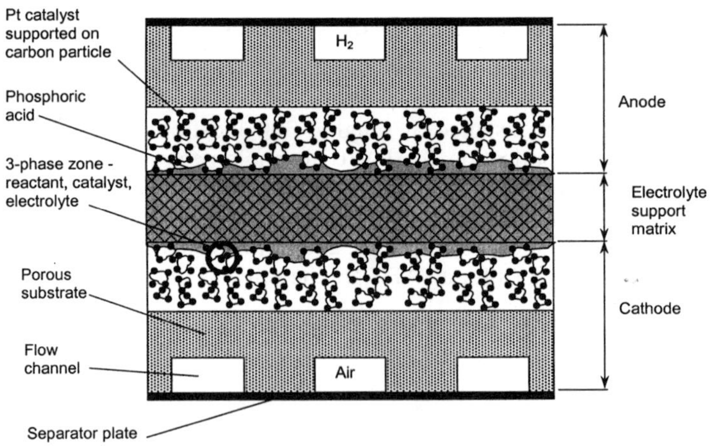

Figure 1-6 PAFC structure.

sizes range from 50 to 200 kWe, but stacks as large as 1 MWe have been constructed. Large stacks can have a cross-sectional area that is on the order of 1.0 m^2 and may contain as many as 350 cells.

The performance of a PAFC stack degrades with time due to corrosion, catalyst poisoning, and other effects. This results in a gradual reduction in cell voltage that can be as low as 2 mV/1,000h [3]. Ultimately, corrosion and/or loss of performance will require that the stack be replaced. Current development targets for stack life are 40,000 hours with 20,000 hours already demonstrated.

The PAFC stack uses hydrogen as its fuel. Hydrocarbon fuels such as methane, propane, landfill gas, and others can be reformed into a hydrogen-rich gas stream that can then be used to fuel the PAFC. Gases other than hydrogen act as diluents or contaminants in the fuel stream. In particular, carbon monoxide and compounds of sulfur can damage the platinum catalysts and must be reduced to acceptable limits in the fuel stream.

Representative performance measures for a PAFC stack are:

Cell voltage: 0.6 to 0.8 V/cell

Current density: 100 to 400 mA/cm^2

Power density: 0.1 to 0.3 W/cm^2

MOLTEN CARBONATE FUEL CELLS (MCFC)

Molten carbonate fuel cells operate at a temperature of approximately 650°C (1200°F) that is much higher than the operating temperature for PEMFC or PAFC technology. At this high temperature, less expensive catalysts may be used, catalyst poisoning is less likely, heat recovery for fuel processing and cogeneration systems is facilitated, and fuel reforming within the stack is possible. These advantages can lead to higher overall system efficiency and greater fuel flexibility. In fact, one of the promising applications for MCFC technology is the production of electricity from gasified coal. Unfortunately, the higher temperatures in conjunction with the corrosive environment in the cell make it a challenge to attain an adequate operating life.

The cell reactions for a MCFC are very different from those for the proton exchange membrane and phosphoric acid cells in that they involve the transport of a negative carbonate ion instead of a proton:

Anode: $$H_2 + CO_3^{2-} \rightarrow H_2O + CO_2 + 2e^- \quad (1\text{-}37)$$

$$\text{Cathode:} \quad \frac{1}{2}O_2 + CO_2 + 2e^- \rightarrow CO_3^{2-} \qquad (1\text{-}38)$$

$$\text{Net:} \quad H_2 + \frac{1}{2}O_2 + CO_2 \text{ (cathode)} \rightarrow H_2O + CO_2 \text{ (anode)} \qquad (1\text{-}39)$$

As shown by Equation 1-38, CO_2 is consumed at the cathode and must be either supplied externally or recycled from the anode. Also, in contrast to the PEMFC and PAFC, the exhaust gas, which consists of H_2O and CO_2, exits from the anode instead of the cathode.

The MCFC usually operates at a temperature of 650ºC (1200ºF) and at pressures ranging from 1 to 10 atm. At this elevated temperature, precious metal catalysts are not required and nickel alloys are typically used at the anode while nickel oxides are used at the cathode. The higher operating temperature also permits the use of CO as an indirect fuel for the MCFC. When CO is supplied with the fuel stream at the anode, it reacts with the water produced at the anode according to the shift reaction:

$$CO + H_2O \leftrightarrow CO_2 + H_2 \qquad (1\text{-}40)$$

The H_2 produced by the shift reaction is then consumed by the cell reaction. This facilitates the operation of the MCFC on reformate fuels including gasified coal.

The cell structure for a typical MCFC is illustrated in Figure 1-7. The anode is constructed of a porous nickel alloy. Fuel gas enters the anode and reacts with carbonate ions from the electrolyte to yield electrons, which travel through the external circuit, and product gases including water and carbon dioxide. Carbon monoxide at the anode can be used to produce additional H_2 through the shift reaction (Equation [1-40]). The anode section of the cell can also be used for internal reforming of simple hydrocarbon fuels as discussed below. The electrolyte is composed of a mixture of Li_2CO_3, Na_2CO_3, and/or K_2CO_3. The electrolyte is retained by capillary pressure in a porous, semisolid $LiAlO_2$ support. The capillary pressure in the small electrolyte support pores causes the support to remain full of electrolyte while the electrodes with larger pore size distributions remain only partially filled with electrolyte. A "bubble pressure barrier" consisting of a fine pore size metal or ceramic material is usually located between the anode and electrolyte matrix. This barrier, due to its fine pore size, retains electrolyte, thus helping to

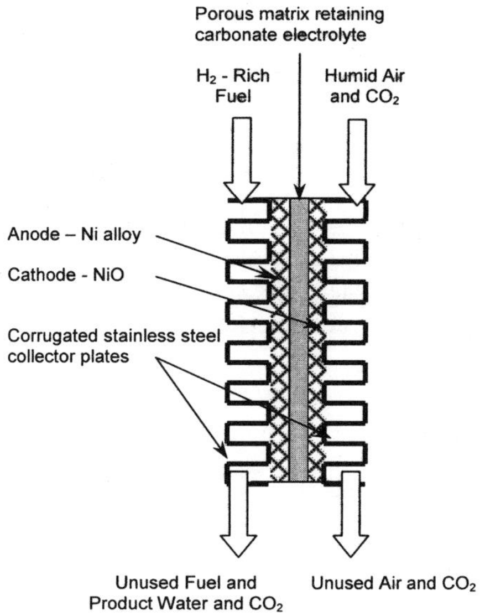

Figure 1-7 MCFC cell structure.

stabilize the electrolyte matrix and forming a barrier to prevent gas crossover [4]. The cathode is constructed of porous NiO. At the cathode, O_2 reacts with CO_2 and electrons from the external circuit to yield carbonate ions that travel through the electrolyte. Unfortunately, the NiO cathode can also react with CO_2, causing Ni from the cathode to dissolve into the electrolyte. This leads to degradation of the cathode and short circuiting of the cell when the Ni precipitates inside the electrolyte. Many of the major developmental challenges for MCFC systems are at the individual cell level and relate to the structural stability of the anode and cathode, dissolution of the cathode, and retention of the electrolyte.

Individual cells are assembled into a stack as illustrated in Figure 1-8. Bipolar plates are located between cells. The bipolar plate serves as the anode collector plate for one cell and the cathode collector plate for the adjacent cell, effectively connecting the cells in series. Cooling is accomplished by recycling the anode or cathode gas and providing a gas cooler in the loop. Stacks as large as 250 kWe have been constructed. Large stacks

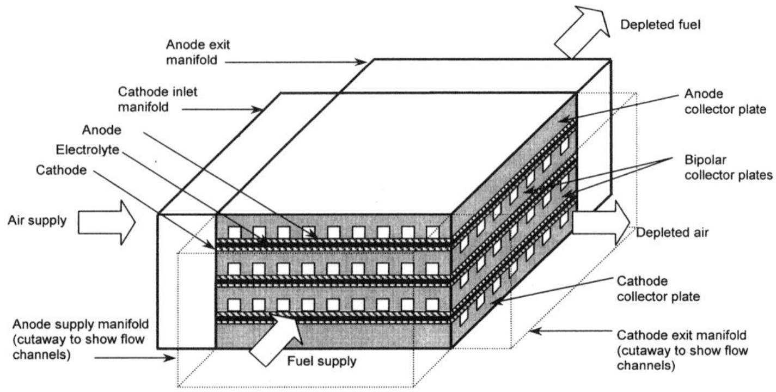

Figure 1-8 MCFC stack configuration with external manifolds.

can have a cross-sectional area that is on the order of 1.0 m^2 and may contain as many as 250 cells.

Although the MCFC can use both H_2 and CO as fuel, hydrocarbon fuels such as methane and natural gas must be reformed to yield a gas stream that is rich in H_2 and CO. With MCFC technology, there are three approaches to fuel reformation, as illustrated in Figure 1-9.

With external steam-reforming (EXR) of methane, the methane and steam are combined and heated in the presence of a catalyst in a separate reactor to yield H_2 and CO according to the steam-reforming reaction:

$$CH_4 + H_2O \leftrightarrow 3H_2 + CO \tag{1-41}$$

The H_2 and CO then become fuel for the MCFC. Heat from the fuel cell can be recovered to provide some of the energy required to heat the methane and steam. Another alternative is to reform the fuel within the fuel cell. This can be accomplished by introducing a steam-reforming catalyst within the anode section (direct internal reforming [DIR]) or by introducing the steam-reforming catalyst in a separate section that exchanges heat with the anode (indirect internal reforming [IIR]). With IIR, the reforming process is the same as with external reforming, but heat transfer is simplified. With DIR, the steam-reforming reaction given by Equation 1-41 is combined with the basic anode reaction given by Equation 1-37 and the water gas shift reaction in Equation 1-40 to yield the net anode reaction:

$$CH_4 + 4CO_3^{2-} \rightarrow 2H_2O + 5CO_2 + 8e- \tag{1-42}$$

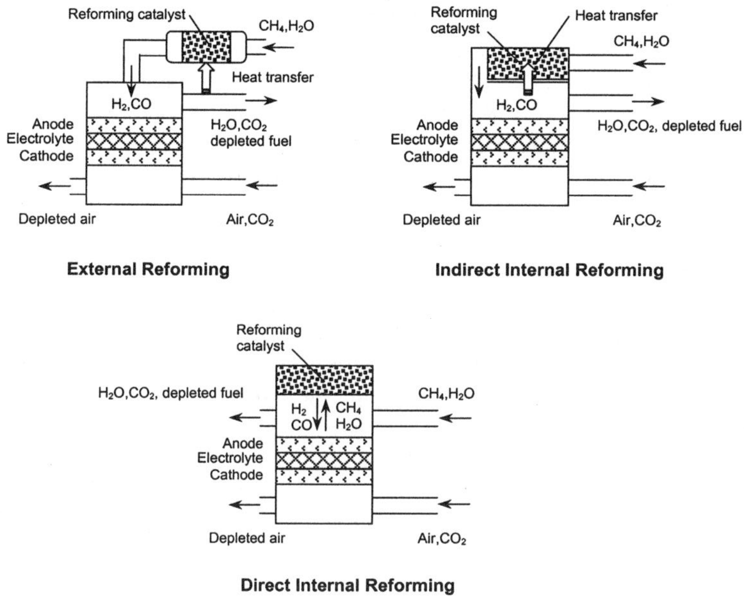

Figure 1-9 Fuel reformation with MCFC systems (adapted from reference [4]).

With direct internal reforming, the cell reaction produces H_2O and consumes H_2, thus favoring the production of H_2 and CO in the reforming reaction and the production of H_2 in the water gas shift reaction. In both IIR and DIR, the energy required by the endothermic steam reforming reaction can be provided by the cell reaction, thus eliminating the need for external heat transfer to the reformer and reducing the heat load on the stack cooling system.

The performance of an MCFC depends on the operating pressure, temperature, and reactant gas composition. The performance of an MCFC increases as the operating pressure increases and increases at a diminishing rate as the temperature increases. Representative performance measures for an MCFC stack operating at 650°C (1200°F) and 10 atm on air and reformate gas from an external reformer (EXR) and for an MCFC stack operating at 650°C (1,200°F) and 1 atm on air and methane reformed directly within the cell (DIR) are as follows:

	EXR	DIR
Cell voltage:	0.9 to 0.85 V/cell	0.95 to 0.80 V/cell
Current density:	50 to 250 mA/cm^2	60 to 140 mA/cm^2
Power density:	0.05 to 0.2 W/cm^2	0.06 to 0.1 W/cm^2

SOLID OXIDE FUEL CELLS (SOFC)

Solid oxide fuel cells operate at the highest temperature of all of the fuel cell technologies. The operating temperature range of 800°C to 1,000°C (1,500°F to 1,800°F) eliminates the need for precious metal catalysts, allows internal reforming of hydrocarbons, and results in high-quality thermal energy for cogeneration or bottoming cycles. In addition, the SOFC offers the advantage of a solid electrolyte, which eliminates the corrosion problems associated with the PAFC and the MCFC and the water management problems of the PEMFC. The challenges of SOFC development include the need to develop inexpensive, easily fabricated materials that can withstand the thermal stress associated with the high operating temperatures.

Like the MCFC, the cell reactions for a SOFC involve the transport of a negative ion, in this case, O^{2-}

Anode: $$H_2 + O^{2-} \rightarrow H_2O + 2e^- \qquad (1\text{-}43)$$

Cathode: $$\frac{1}{2}O_2 + 2e^- \rightarrow O^{2-} \qquad (1\text{-}44)$$

Net: $$H_2 + \frac{1}{2}O_2 \rightarrow H_2O \qquad (1\text{-}45)$$

In addition, like the MCFC, the SOFC can use CO as a fuel through the water gas shift reaction (Equation 1-40) or by direct oxidation. Methane can be reformed within the SOFC by the steam reforming reaction (Equation 1-41) to yield H_2 and CO, which are then utilized through the water gas shift or cell reaction.

The use of solid phase components permits a wide variety of cell configurations including tubular, segmented tube, flat plate, and monolithic designs. The tubular cell structure, illustrated in Figure 1-10, is at the most advanced state of development. The anode is constructed of a cermet of nickel and a porous ceramic structure composed of yttria (Y_2O_3) stabilized zirconia (ZrO_2). The anode surrounds the electrolyte and the cathode. Fuel enters from the perimeter of the anode and reacts with O^{2-} ions from the electrolyte to yield electrons that travel through the external circuit and product gases including water and CO_2 (if CO is part of the fuel stream). The electrolyte consists of yttria-stabilized zirconia. The presence of the yttria increases the number of vacant sites in the zirconia lattice that permit the movement of O^{2-} ions at high temperatures. The electrolyte surrounds and is supported by the cathode, which is composed of lanthanum manganite. Oxygen from air inside the cath-

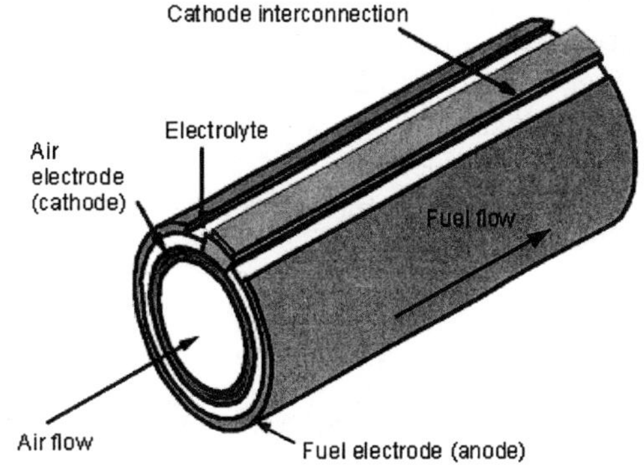

Figure 1-10 Tubular SOFC structure.

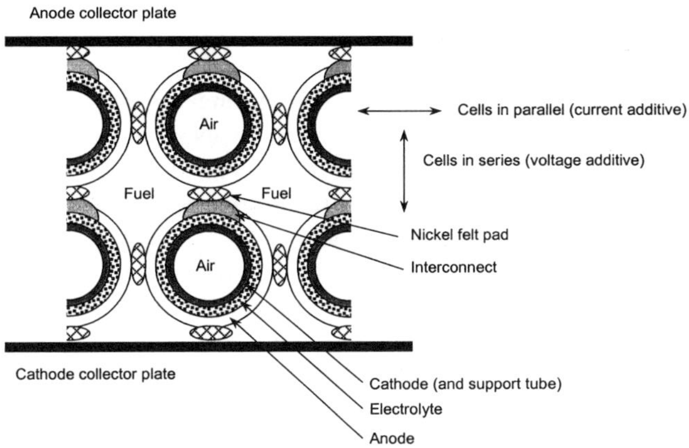

Figure 1-11 SOFC stack configuration.

ode tube reacts with electrons from the external circuit to produce the O^{2-} ions that travel through the electrolyte. An interconnect element penetrates the anode layer (without electrical contact) and the electrolyte layer to provide an electrical connection to the cathode.

Individual tubular cells are assembled into a stack as illustrated in Figure 1-11. In this configuration, fuel enters from the chamber surrounding the tubes. Air is introduced through pipes that extend into the SOFC tube. Cells are arranged in a matrix and connected electrically such that rows of cells are connected in parallel while cell columns and are connected in series. Electrical connections between cells are made through nickel felt pads. Heat from the cell reaction is transferred to the fuel and air inlet streams, thus eliminating the need for a separate cooling system. Typical cells are 1 to 2 cm in diameter and range from 30 to 150 cm in length. Stacks as large as 250 kWe, containing more than 1,000 cells have been constructed.

Representative performance measures for a SOFC stack are

Cell voltage: 0.6 to 0.8 V/cell

Current density: 100 to 400 mA/cm^2

Power density: 0.1 to 0.3 W/cm^2

SUMMARY OF FUEL CELL CHARACTERISTICS

Table 1-1 summarizes the important characteristics of each of the fuel cell technologies. Information on fuel cell *system* cost and performance is presented in Chapter 4 following a discussion of how fuel cells are integrated into fuel cell systems.

FUEL CELL DEVELOPMENT

Research and development is currently underway for all four fuel cell technologies. In general, efforts are focusing on reducing cost, increasing stack life, and integrating cell stacks into fuel cell systems. Chapter 2 provides additional details regarding each technology and the research, development, and commercialization efforts currently in progress. Chapter 3 describes how fuel cell stacks can be integrated into fuel cell systems that supply electricity and heat. Chapter 4 discusses the issues associated with incorporating fuel cell systems into building applications.

TABLE 1-1:
Summary of Characteristics of the Major Fuel Cell Types

Fuel Cell Technology	PEMFC	PAFC	MCFC	SOFC
Electrolyte	Polymer Membrane	Phosphoric Acid	Molten Carbonate	Ion-Conducting Ceramic
Operating Temperature	50°C to 90°C 120°F to 190°F	200°C 390°F	650°C 1,200°F	800°C to 1,000°C 1,500°F to 1,800°F
Charge Carrier	H^+	H^+	$CO_3^=$	$O^=$
Anode	Platinum on carbon	Platinum on carbon	Nickel	Nickel/zirconia cermet
Cathode	Platinum on carbon	Platinum on carbon	Nickel Nickel oxide	Ion/electron conducting ceramics
Primary Stack Components	Carbon-based	Carbon-based	Stainless Steel	Ceramic
Ext. Reformer Reqd. for CH_4	Yes	Yes	No	No
Stack Water Removal	Liquid/vapor in cathode exhaust	Vapor in cathode exhaust	Vapor in anode exhaust	Vapor in anode exhaust
Stack Heat Removal	Reactant gas and coolant	Reactant gas and coolant	Reactant gas and internal reforming	Reactant gas and internal reforming
Power Density (W/cm^2)	0.3 to 0.8	0.1 to 0.3	0.1 to 0.2	0.1 to 0.3

Chapter 2

FUEL CELL TECHNOLOGY SURVEY

LITERATURE REVIEW

The following sections present the status as of 2000 of each of the fuel cell types introduced in the previous chapter. Each section is divided into subsections describing the materials of construction, the manufacturing, and the operational characteristics of each technology as found in the recent literature. The major commercialization efforts, the proposed applications, the latest demonstration projects, and the challenges to future development are also discussed.

PROTON EXCHANGE MEMBRANE FUEL CELL (PEMFC)

The proton exchange membrane fuel cell (PEMFC) is the most promising and rapidly advancing fuel cell technology for transportation and small to mid-size stationary applications. It was first introduced in the late 1960s with the Gemini Space Program, and was neglected thereafter until the 1980s due to high costs resulting from high loadings of precious metal catalyst, intolerance to impurities in the fuel and oxidizer streams, and low power densities. Major advances in performance and cost reduction have been achieved during the last decade. However, additional improvements are required before the PEMFC can compete with technologies such as the electricity grid in stationary applications and IC engines in transportation applications. The following discussion emphasizes the recent developments (1990 to 2000) in PEMFC technology, including advancements in materials, improvements in operation and design principles, cost reduction efforts in manufacturing, and commercialization efforts. The information presented is from the open literature and indicates general trends and research areas. Specific details are sometimes unavailable or delayed in publication because companies consider the information proprietary.

Performance and Operating Conditions

The proton exchange membrane fuel cell is operated at temperatures in the range of 50°C to 90°C (120°F to 190°F) and at pressures of 1 to 3 atm. To maximize the power output of the fuel cell, the polarization curve must be kept flat and the limiting current density extended as far as possible. This goal is pursued through development of improved materials and better design of the fuel cell assembly. A number of studies have examined the influence of specific components of the fuel cell (e.g., membrane electrode assembly) in an effort to understand and improve the behavior of the cell [6-13].

The operating conditions of the fuel cell have a strong effect on its performance. Mathematical models have been proposed to describe the entire cell, including flow channels and bipolar plates, and to describe overall stack behavior. These models permit evaluation of different operating conditions and determination of the optimum combination of operating parameters to maximize efficiency and power density of the PEMFC [14-20].

The performance of the cell also depends on the properties of the fuel and oxidizer streams. Reformate fuels derived from hydrocarbon fuel sources contain various species other than H_2 that can affect performance. For example, if the CO concentration in the fuel stream exceeds 10 ppm, catalyst poisoning and severe performance degradation are experienced. Table 2-1 shows the effect of various fuel gas species on fuel cell performance.

Shishtla et al. [21] experimentally evaluate the effect of fuel composition and other operating conditions including temperature, current density, and fuel utilization on fuel cell performance and endurance. Their results show cell degradation of less than 1%/1,000 h over a 5,000 h period for synthetic reformate free of CO.

TABLE 2-1:
Effect of Fuel Gas Constituents on Fuel Cell Performance [22]

Gas Species	Effect
H_2	Fuel
CO	Poison (> 10 ppm)
CH_4	Diluent
CO_2 and H_2O	Diluent
Sulfur compounds (e.g., H_2S)	No studies found

Materials of Construction and Manufacturing

Polymer Membrane Electrolyte. PEMFCs utilize a sulfonic acid polymer membrane, typically 25 to 100 μm thick, as the electrolyte. A number of companies, including DuPont, Dow Chemical Company, Gore, and Ballard Advanced Materials Corporation, are working to develop membranes with lower cost, improved protonic conductivity, and longer life [23, 24]. Research is also underway to develop membranes that are suitable for operation at higher temperatures.

The membrane must remain hydrated to permit the transport of protons. However, the electrode surfaces must be kept free of liquid water, which blocks the flow of fuel and oxidant to the reaction sites. Most of the current fuel cell designs require that the inlet gases be humidified to a high relative humidity while avoiding condensation. An alternative to humidifying the inlet gases could be the development of self-humidifying polymer electrolyte membranes. Self-humidification allows the use of very thin membranes, improves cell performance, simplifies the auxiliary systems, reduces cold start problems, and facilitates proper fuel cell operation with abrupt load changes. The membrane can be made self-humidifying by incorporating oxide materials that absorb water. Modifying the membrane to make it self-humidifying leads to a small loss of hydrogen, which is estimated to be less than a few percent under practical operating conditions [25].

Electrodes and Catalyst Layer. Research and development efforts for the electrode and catalyst layer focus on extending the life of the cell, reducing the amount of catalyst per active area to lower the cost, and improving the tolerance of the catalyst to carbon monoxide. Platinum (Pt) based alloys are the most commonly used catalysts for PEMFCs. During the last decade, the amount of platinum required at the electrodes has decreased from 4 mg/cm^2 to 0.6 mg/cm^2 at the cathode and less than 0.25 mg/cm^2 at the anode. Recently, an experimental PEMFC stack using less than 1.5 g/kWe platinum was operated on pure hydrogen at 0.7 V and 0.4 W/cm^2 without any significant performance drop during 3,000 hours of continuous operation [26]. Sirinivasan et al. describe methods for achieving high-power density with low catalyst loading [27]. Table 2-2 describes the improvements of the catalyst layer during the period 1990-1997 and the corresponding reduction in cost of the electrodes [28].

**TABLE 2-2:
Development of Pt-Loading and Specific
Electrode Material Costs for PEMFCs [28]**

	1990	1997	Long-Term Goal
Pt-loading anode (mg/cm^2)	4	0.2	0.1
Pt-loading cathode (mg/cm^2)	4	~2	0.2
Total Pt-loading (mg/cm^2)	8	~2.2	0.3
Maximum power (W/cm^2)	0.5	0.8	1.0
Pt-electrode cost ($/kWe) (1 g Pt = $11.50)	184	~32	3.50

In addition, researchers are working to develop improved catalyst materials. Denis et al. have suggested that the use of ball milled Pt and Pt/Ru catalysts can improve CO tolerance at the expense of a reduction in the specific area of the catalyst [29]. Oetjen et al. have also observed improved CO tolerance with Pt-Ru catalysts [30]. Swette et al. evaluated possible catalyst materials for regenerative PEMFCs. They note that the RuO_X catalyst is a promising alternative to the commonly used Pt and Pt-Ir catalysts for various PEM applications [31].

Another approach to improving CO tolerance would be to develop a method for oxidizing the CO. One solution would be to introduce air into the fuel stream, allowing direct contact of oxygen and carbon monoxide on the fuel side of the stack. Another solution is to use high-frequency bursts of current at the anode side of the cell to oxidize residual carbon monoxide molecules attached to the catalyst molecules. The suggested method consumes 5% to 10% of the electrical power generated in the stack. [32].

Manufacturing. The success of PEMFC technology will depend not only on improved performance but also on reduced production costs. Manufacturers of PEMFCs are developing improved manufacturing techniques that will facilitate mass production.

In a typical PEMFC stack, the bipolar plates constitute over 80% of the mass and almost all of the volume. The three common processes to produce the bipolar plate are

- machining the desired patterns,
- injection molding a composite carbon part followed by graphitization at temperatures exceeding 2,500°C, and
- injection molding using a graphite-filled polymer.

Murphy et al. propose an inexpensive way of producing lightweight bipolar plates for PEMFCs that uses metals at all locations where high conductivity is required and engineering thermoplastics fabricated by injection molding at other areas. The material cost of the stack and its weight to volume ratio are decreased [33].

Demonstration and Commercialization Projects

The PEMFC can be used for a variety of applications, including combined heat and power generation at stationary plants (in the range of 1 kWe to 1 MWe), power generation in transportation (in the range of 30 kWe to 300 kWe), and other smaller capacity applications. There are many companies involved in the development of PEMFC systems. The following summarizes some of the larger efforts for both stationary and transportation applications. A more comprehensive list is available from the sources listed in Appendix A.

Stationary Applications.

Analytic/American Power. Together these organizations are developing fuel processors, fuel cell stacks, and integrated fuel cell systems. Analytic Power is developing a residential fuel cell system with a peak capacity of 10 kWe and a continuous capacity of 3 kW. Projected specifications include a conversion efficiency of 40%, cogeneration efficiency of 80% with waste heat available at 60°C (140°F), and a system volume of 0.8 m^3 (9 ft^3). Estimated cost ranges from $4,000 per unit at a production volume of 10,000 units to $3,000 per unit at a production volume of 100,000 units [34].

Avista Laboratories. Avista is developing a modular, low-pressure, air-cooled PEM stack with a 720 W capacity for residential power applications.

Ballard Power Systems Incorporated. Ballard, together with Mitsubishi, has developed a 5 kWe PEMFC operating on natural gas. The power is sufficient to cover the average power requirements of a typical household. Ballard has also developed 7 kWe and 30 kWe systems operating on hydrogen and a 10 kWe system operating on natural gas. Ballard plans to commercialize 250 kWe fuel cell systems operating on natural gas with electrical conversion efficiencies of 40% and cogeneration efficiencies of 80% with thermal energy available at 80ºC (170ºF) [34]. This product is designed as primary or backup power for commercial and industrial buildings such as hotels, hospitals, manufacturing facilities, and retail shopping centers. Ballard has begun field demonstrations of prototype 250 kWe power plants operating on natural gas [35, 37]. Another Ballard 250 kWe PEMFC system with a diesel fuel processor is being tested at the Royal Military College in Kingston, Canada. This system is one of the first demonstrations of combined heat and power generation in larger scale plants using PEMFC technology. The diesel fuel is vaporized with excess steam before entering the reformer, which operates at 600ºC (1,100ºF). The fuel cell stack operates at a temperature of 80°C (175ºF), a pressure of 3 atm (30 psig), and a current density of 400 mA/cm^2. The gross electric efficiency for the system is 46%. The electrical conversion efficiency, including parasitic energy losses is calculated to be 38.6%. The system also produces excess heat in the amount of 230 kWe that can be used for thermal loads such as space or domestic water heating [38].

H Power. H Power is developing stationary fuel cell systems in the 2 to 50 kWe range with plans for systems as large as 250 kWe. Their initial residential unit will have a continuous power output of 3 to 4 kWe and will include a fuel processor for natural gas or propane and a battery to provide a peak power of up to 10 kWe. The system is expected to produce hot water for cogeneration at a temperature of 55ºC (130ºF). The system is expected to be commercially available by 2001 with near-term cost of $5,000/kWe and long-term costs of $1,500/kWe [34].

Plug-Power Incorporated. On June 17, 1998, Plug Power demonstrated its proprietary Plug Power 7000, a 7 kWe residential power system, at a home in upstate New York. Since then, the system has operated on a continuous basis. In August 1999, the system was converted to natural gas operation from pure hydrogen. Plug Power plans to introduce several different commercial models with 40% electrical efficiency that can operate on natural gas, propane, or methanol. The cogeneration efficiency can exceed 80% by using the excess heat generated by the fuel cell for space and water heating [39].

Transportation Applications.

USCAR Partnership. Daimler-Chrysler, Ford, and General Motors founded the partnership USCAR (United States Council for Automotive Research) in 1993 to work together on shared technological and environmental concerns. One of their focuses is research and development of PEMFCs for transportation applications. One of the significant accomplishments under the partnership is the demonstration of a fuel cell system using gasoline as the fuel. The goal of the partnership is the development of high-mileage, fuel-flexible, and low-emission electric vehicles that can be conveniently refueled at existing gas stations [40].

U.S. Department of Energy (DOE), Arthur D. Little, Plug Power Inc., Los Alamos National Laboratory, and Ballard Power Systems completed a cost-shared five-year program that successfully demonstrated fuel cell systems that can generate electricity from a variety of fuels, including gasoline, ethanol, methanol, and natural gas. When running on gasoline, a fuel cell passenger car is expected to be many times cleaner than the Tier 2 emissions standards of the U.S. Environmental Protection Agency. Project partners proposed developing a full-scale fuel cell, a fuel processor, and high-volume manufacturing technologies to reduce cost. The fuel cell systems will be installed and operated in an automobile to demonstrate fuel economy, practicality, performance, reliability, and durability [40].

U.S. Department of Energy's Office of Transportation Technologies, General Motors Research Center, Delphi Energy and Engine, E.I. DuPont de Nemoirs, and Ballard Power Systems. The U.S. Department of Energy is sponsoring a proof-of-concept 30 kWe PEMFC power generator with a methanol fuel processor. In addition to meeting power requirements and exceeding ultra low emission vehicle (ULEV) standards, the system is expected to demonstrate vehicle packaging viability and the cost-effective nature of the technology. The achievements of the project to date are the integration of a methanol fuel processor with a 50 kWe PEMFC stack and its operation as a system. The contribution of General Motors has been the development of a 25 kWe stack, which features sensors and controls for automated operation. Computer-based simulation models were developed and are guiding continual improvements in system efficiency. Ballard has delivered a 50 kWe fuel stack, while Delphi built a 30 kWe methanol processor that will be used in the project [40].

Ballard Power Systems Incorporated. Ballard is supplying fuel cells and fuel cell systems to major companies and organizations, including GM, Ford, Daimler-Chrysler, Chicago Transit Authority, Volkswagen, Translink, Honda, and Nissan for testing purposes. The efforts in commercializing the PEMFC at Ballard are focused on public transit and automobile applications [35]. For public transportation, Ballard has developed a 275 HP fuel cell engine for a hydrogen-fueled 40-ft heavy-duty transit bus. Following the completion of this project, the Chicago Transit Authority and British Columbia Transit each agreed to acquire three hydrogen-fueled prototype buses. In a two-year test period, these test fleets will provide performance, cost, and reliability data. Concurrent with this phase, the commercial prototype engine is being designed and built and will also undergo extensive field testing to refine the design and meet commercial engine expectations. The current range and capacity of the bus is 250 miles and 60 passengers, while the goal for 2002 is a range of 350 miles and a capacity of 75 passengers. For automobiles, developments at Ballard have included a 50 kWe methanol-fueled PEMFC system with a power density of 1000 W/L [36, 40, 41, 42]. Current efforts are focused on demonstrating the practicality of fuel cells in concept cars and sport utility vehicles and on commercialization issues such as cost reduction and development of high volume manufacturing processes [42].

Challenges

The recent improvements in cost and performance in PEMFCs allow them to be considered as alternative power sources with high fuel efficiencies and low emission characteristics. However, before PEMFCs can be commercialized, there are still many aspects of the technology that need further improvement and refinement. The main challenges to the development of PEMFC technology are

- reduction of the system costs to the level of the competing technologies (e.g., internal combustion engines), meaning a decrease of at least two orders of magnitude from the present costs [28, 43, 44],
- development of a better understanding of the influence of operating conditions including reformate composition, reactant humidification, stack cooling, etc.
- development of stack designs and manufacturing processes for economical high volume production [33],
- development of improved cold-start capability and better response to transient loads,
- development of efficient low-cost systems for air compression and humidification,

- development of efficient, compact, inexpensive fuel processors [44],
- better tolerance to catalyst poisoning, and
- reduction of the catalyst loading [28].

Economics

The following discussion of the economics of PEMFCs is divided into two separate sections according to the main application types. Although identical PEMFC stacks are used in stationary and transportation applications, the goals of the programs and costs of the systems are considered separately.

Stationary (Lifetime Based on 40,000 Operating Hours). To be competitive with the utility grid, fuel cell electricity costs must be reduced to the range of $0.06/kWh to $0.08/kWh. In general, the cost of electricity supplied by a fuel cell system can be calculated by considering the capital cost, operating and maintenance costs, and the fuel cost. Fuel costs must include the cost of the source fuel and equipment required to reform the fuel. Barbir and Gomez present a study of PEMFC economics based on a theoretical fuel cell operating on pure hydrogen. With current PEM technology (at an assumed stack cost of roughly $1,900/kW) and a hudrogen cost of $20/GJ, the electricity cost is estimated to be $0.24 to $0.60 per kWh, depending on the operating conditions of a 10 kWe fuel cell stack. With expected developments in the performance of the fuel cell and reductions in the cost of the cell, stack, and the fuel, electricity cost is projected to be as low as $0.08 to $0.12 per kWh, without including heat recovery from the system [43]. If the fuel cell is used in a residential application, cogeneration will decrease the actual power requirements of the building as the space heating and warm water requirements can be provided by waste heat from the stack.

Transportation (Lifetime Based on 5,000 Operating Hours). The goal for PEMFCs in transportation applications is to reduce the cost of the system to the level of the internal combustion engine, which typically costs less than $50/kWe. For this purpose, the target values, tabulated in Table 2-3, are set for the near future of PEMFC research and development.

PHOSPHORIC ACID FUEL CELL (PAFC)

The phosphoric acid fuel cell was the first type of fuel cell to be commercially available. Research and development work for PAFC fuel cell systems has been in progress for the last two decades. Fuel cells

TABLE 2-3:
Program for New Generation Vehicle Targets [44]

Year	2000	2004
Fuel cell stack ($/kWe)	100	35
Fuel processor ($/kWe)	30	10
Total fuel cell system ($/kWe)	150	50

based on PAFC technology are operating at over 300 sites worldwide. Since 1994, the ONSI Corporation has manufactured the PC25 200 kWe system, which is installed at more than 200 sites. Field experience has led to further improvement in system life, reliability, and cost. The performance, ongoing research activities, recent commercialization efforts, and challenges associated with the development of PAFC technology are described in the following sections.

Performance and Operating Conditions

PAFCs typically operate at temperatures of approximately 200°C (390°F), in the voltage range of 0.60 to 0.80 V, and at current densities of 100 to 400 mA/cm^2. The power density varies between 0.1 to 0.3 W/cm^2. Increases in operating pressure and temperature are known to enhance the reaction rates at the electrodes and the performance of the PAFC. For example, the ONSI Corporation has developed an atmospheric PAFC operating at approximately 205°C (400°F), which provides 0.75 V, 240 mA/cm^2, and 0.180 W/cm^2. In contrast, ONSI's pressurized (8.2 atm) design yields a performance of 0.71 V, 431 mA/cm^2, and 0.307 W/cm^2 [22]. Operating temperatures are limited because high temperatures result in faster catalyst sintering, material and component degradation, corrosion, and electrolyte loss through evaporation [22].

In addition to temperature, the operating voltage affects the rate of degradation of the cell. Aragane and Urushibata estimated the effect of the operating voltage on the cell voltage decay rate in a phosphoric acid fuel cell. Their findings indicate that the cell voltage decay can be estimated as a function of the operating voltage and temperature and that the decay can be limited to 1 mV per 1000 hours by operation at voltages below 0.84 V and temperatures below 200°C (390°F) [45].

With proper fuel processing, PAFCs can operate with a variety of fuel gases, including natural gas, propane, and landfill gas. However, the fuel processor must limit the concentration of certain gases in the

TABLE 2-4:
Effect of Fuel Gas Constituents on Fuel Cell Performance [22]

Gas Species	Effect
H_2	Fuel
CO	Poison (> 0.5 percent)
CH_4	Diluent
CO_2 and H_2O	Diluent
Sulfur compounds (e.g., H_2S)	Poison (>50 ppm)

fuel stream. The effects of various gases on fuel cell performance are given in Table 2-4.

Materials of Construction and Manufacturing

Electrodes and Electrolyte. The materials commonly used for the manufacturing of PAFC electrodes include a PTFE-bonded carbon paper that supports carbon black particles that are bonded together with PTFE and impregnated with platinum catalyst. Typical catalyst loadings are 0.10 mg Pt/cm^2 at the anode and 0.5 mg Pt/cm^2 at the cathode [5]. The electrolyte is pure phosphoric acid (H_3PO_4) supported in a PTFE-bonded silicon carbide matrix. A comprehensive review of the literature related to PAFC electrode and electrolyte development is provided by Appleby and Foulkes [46]. Current development efforts for electrodes and electrolytes focus on catalyst improvements that reduce sensitivity to catalytic poisoning, determination of optimal catalyst loading, development of advanced gas diffusion electrodes, and improved corrosion tolerance [22].

The two common manufacturing methods for PAFC electrodes are the coating method and the rolling method. In the conventional coating method, an electrode is obtained by coating a slurry of the electrocatalyst on the electrode support. Drying and sintering the electrode completes the process. This method of fabricating the electrodes has the major disadvantage that an abundance of cracks occurs in the catalyst layer during the drying and sintering. An appropriate number of cracks in the catalyst layer may improve the performance of a fuel cell, but a large number causes the phosphoric acid electrolyte to flood in the cracks, greatly deteriorating the performance of the fuel cell. Another important problem involved in this process is that the electrocatalyst film can easily be detached from the support because of the weak attraction forces between the carbon particles within the electrocatalyst layer [5, 47].

In the conventional rolling method for manufacturing an electrode, a gumlike slurry is produced, which is rolled into a sheet type catalyst layer. The catalyst layer is then pressed onto the electrode support. The difficulties involved with this method are the rolling of a wide catalyst layer of uniform thickness and applying a uniform stress over this wide layer during the pressing process. These problems make the process technically difficult and expensive [47].

A new method for the manufacturing of PAFC electrodes has been proposed by Song et al., which is a combination of the coating and rolling methods. In this method, an electrocatalyst slurry is coated upon an electrode support, dried at 225ºC (440ºF), subjected to a rolling process, and then to a sintering process at 350°C (660ºF) in an inert atmosphere. The combination method is shown to have better performance than the coating and rolling methods [47].

Demonstration and Commercialization Projects

The first commercially available fuel cells were PAFCs. During the last decade, many PAFC power plants have been installed and tested worldwide. The capacity of the demonstration plants varies from portable power systems up to an 11 MWe utility scale system. The major developers of PAFC technology include International Fuel Cells (IFC) in the U.S. and Fuji Electric Corporation, Toshiba Corporation, and Mitsubishi Electric Corporation in Japan [22]. Demonstration and commercialization efforts by IFC and Fuji are discussed below.

International Fuel Cells (IFC). The well-known PC25 power generation system developed by ONSI, a subsidiary of IFC, has been available since 1992. The system provides 200kWe electrical power and up to 205 kWt of thermal energy at 140°C (280ºF). The system has been installed at 170 locations—in 25 states in the U.S., seven countries in Europe, four in Asia, and one in Australia. The longest operating time for a unit is 45,000 hours and the longest continuous run is 9,500 hours. The total operating time for all the units is over 3,000,000 hours [48-50]. Nymoen discussed the first results of the PAFC demonstration plants installed in Europe in 1994. Four 50 kWe Fuji systems and nine PC25 systems by ONSI were installed in the area by 1992, and the performance of these systems is evaluated in the report [51].

Fuji Electric Company began installing 27 demonstration plants with 2,100 kWe total capacity in 1992 with plans to install an additional 63 plants ranging from 50 to 5,000 kWe. The company is working on

PAFCs using LPG and methanol fuels along with the more common natural gas-fueled systems. [50, 52].

Together with the Tokyo, Osaka, and Toho Gas companies, Fuji Electric installed eighteen 50 kWe and thirteen 100 kWe PAFCs for testing purposes in 1989 in Japan. In 1995, more than ten units had 10,000 hours cumulative operating time, while tests on cell durability and cost reduction were being conducted for the commercialization of the systems [53].

Challenges

A survey of the PAFC demonstration plants and projection of their results into the future indicate that the following improvements are necessary before the technology can be widely commercialized [51]:

- Verify the durability and reliability for 40,000 hours operation (the initial performance target of 20,000 hours was demonstrated in 1996) [53, 54].
- Reduce manufacturing costs for the PAFC systems without compromising the performance[53].
- Increase the reliability of the system.
- Reduce space requirements of the system without restricting the access to plant components.
- Recover heat at higher temperature levels (to facilitate applications involving absorption chillers and process steam).

In addition, PAFC technology faces competition from the other fuel cell technologies that are likely to offer higher efficiencies, higher power densities, and lower costs as they mature.

Economics

Although the installation cost for the PAFC systems was about $9,000/kWe in the early 1990s, it had dropped significantly to $4,500/kWe in 1994 [51]. Currently, the system is available for roughly $5,000/kWe [48]. To compete with the existing diesel generators, the prices must drop below $1,550/kWe in the U.S. and Europe [55].

MOLTEN CARBONATE FUEL CELL (MCFC)

Research and development of molten carbonate fuel cells (MCFCs) actually started in the 1950s. Since then, great progress has been made with respect to understanding the cell operation and improving the materials, performance, and manufacturing techniques. Currently, demonstration plants in the 250 kWe to 2 MWe range have been developed. Commercialized systems will have capacities as high as 10 MWe. Because of their relatively high operating temperature, the MCFCs are attractive for combined heat and power applications (such as commercial buildings, hospitals, hotels). With current and projected MCFC technologies, 50% to 60% of the HHV of the fuel can be converted into electricity and an additional 20% can be converted into high-quality waste heat [56].

Performance and Operating Conditions

Molten carbonate fuel cells will generally operate at current densities in the range of 100 to 200 mA/cm^2 and at cell voltages of 0.75 to 0.95 V/cell, thus yielding a power density in the range of 0.10 to 0.15 W/cm^2 at operating temperatures of 600°C to 650°C (1,100°F to 1,200°F) [5]. The cell voltage of a MCFC increases with increasing temperature [57] and pressure [58]. However, temperatures in excess of 650°C (1,200°F) lead to increased electrolyte loss and corrosion. In addition, increasing pressure can lead to carbon deposition and methanation in the fuel stream.

MCFCs can operate with a variety of fuel gases including gasified coal. A review of the effect of fuel gas composition on the anode potential is given by Lu and Selman [59]. At the MCFC operating temperature of 650°C (1,200°F), carbon monoxide can react with water to produce additional hydrogen and is thus an indirect fuel for MCFCs. Other gases, however, can reduce performance. Hirshenhofer summarizes the results from several studies of the effects of fuel constituents on MCFCs [22]. The effects of a few of the more common fuel gas species are given in Table 2-5.

TABLE 2-5:
Effect of Fuel Gas Constituents on Fuel Cell Performance [22]

Gas Species	Effect
H_2	Fuel
CO	Fuel
CH_4	Fuel (if internally reforming) or diluent
CO_2 and H_2O	Diluent
Sulfur compounds (e.g., H_2S)	Poison (> 0.5 ppm)

Materials of Construction and Manufacturing

Electrodes and Electrolyte. The evolution of the cell components since 1965 has been discussed in detail by Kordesch and Simader [5]. The commonly used materials for the electrodes and the electrolyte support in MCFCs today are presented below.

Anode:	Ni-10 wt.% Cr, with 50-70% porosity and 3-6 μm pore size, 0.5-1.5 mm thickness with 0.1-1 m^2/g
Cathode:	Lithiated NiO, with 70-80% initial and 60-65% porosity after lithiation and oxidation, 7-15 μm pore size, 0.5-0.75 mm thickness, and 0.5 m^2/g surface area
Electrolyte support:	γ-$LiAlO_2$ ceramic support, 0.1-12 m^2/g with 0.5 mm thickness
Electrolyte:	62 $LiCO_3$-38 K_2CO_3, 50 $LiCO_3$-50 Na_2CO_3, 50 $LiCO_3$-38 K_2CO_3 ~50wt % in tape cast form with 0.5 mm thickness

Development of a long-lasting electrode/electrolyte assembly is complicated by a number of challenges including structural stability, NiO dissolution, and electrolyte loss. The NiCr anode material is subject to creep when the stack is assembled and compressed. Other materials, including copper-based alloys, are under development for use in the anode [60]. At the cathode, the NiO tends to dissolve into the carbonate electrolyte and migrate toward the anode, resulting in dissolution of the cathode and formation of conductive structures that short out the cell. Possible solutions to this problem include alternative cathode materials, thicker electrolytes, and changes in the electrolyte composition [22]. In particular, $LiCoO_2$ as a cathode material seems to be a better selection because it does not precipitate into the electrolyte, is less affected by operating conditions, and has a dissolution rate that is one order of magnitude less than the standard cathode materials. However, the cost of cobalt is higher than nickel [61]. Processes for fabricating electrolytes include hot pressing and tape casting. Separator plates are typically fabricated from thin corrosion-resistant steel alloy (e.g., 316 stainless steel) sheets coated with Ni on the anode side to reduce corrosion [62].

Stacks and Systems. The MCFC operates at a sufficiently high temperature to allow methane to be reformed over a catalyst. By incorporating the reforming catalyst in the stack, preferably in the anode chamber of the individual cells (direct internal reforming—DIR), the heat for the endothermal reforming reaction is provided by the exothermal cell reaction. DIR reduces the heat and mass transport to a minimum. The drawbacks of this design, however, are the increased stack cost due to the incorporation of the reforming catalyst and the possible poisoning of the catalyst due to the highly corrosive molten carbonate electrolyte. An alternative to DIR is indirect internal reforming (IIR) where the catalyst is incorporated in the stack in separate compartments [63]. IIR results in a more complicated stack design and also lower efficiencies. For advanced systems with sizes of 250 kWe to several MWe, internal reforming may reduce the system cost by 50 percent or more, compared with external reforming systems [62].

Another criterion that affects the system performance is the networking of the stacks. The cooling demands for the system can be significantly decreased by connecting smaller-sized stacks in series and injecting relatively cold process gases between the stacks instead of forming one line of parallel-connected cells. The alternative design would result in less parasitic power consumption, lower cooling demands, 10% higher overall efficiency, replacement of high-temperature valves with low-temperature stack control valves, and a substantial reduction in the cost of the stack [62].

Demonstration and Commercialization Projects

Currently, the major developers of MCFC technology include Fuel Cell Energy (formerly Energy Research Corporation) in the U.S.; Brandstofel Nederland (BCN), MTU Friedrichshafen (Germany), and Ansaldo (Italy) in Europe; and Hitachi, Ishikawajima-Harima Heavy Industries, Mitsubishi Electric Corporation, and Toshiba Corporation in Japan [1].

FuelCell Energy. The FuelCell Energy technology uses direct internal reforming technology. In 1993, the company demonstrated a 120 kWe prototype for a natural-gas-fueled power generation system. The actual 2 MWe power generation system was first operated in 1996 in Santa Clara, California [5]. FuelCell Energy currently plans to introduce a product line by 2005 with units ranging from 250 kWe to 3 Mwe. FuelCell Energy has also proposed hybrid cycles that would achieve efficiencies as high as 70% to 73% by using the waste heat from the fuel cell to drive either gas or steam turbines [34].

International Efforts. In Japan, development efforts for MCFCs are focused on the development of large-scale stack technology, balance-of-plant equipment, fuel gas-related technology for gasification, and total systems. Their goal in 1998 was to develop a series of 10 kWe systems with a 40,000-hour lifetime by 1999 and continue research to improve the performance and reliability of the stacks and system and to reduce cost. Their proof-of-concept design will be demonstrated by a 1,000 kWe pilot plant [5, 56].

According to Kordesch, Ansaldo in Italy was building an automated fabrication facility to manufacture MCFC systems up to 8 MWe/year in joint development programs with other European companies [5]. In Germany, MTU Friedrichshafen (now part of Daimler Chrysler) has developed a 260 kWe system that is designed for commercial and industrial cogeneration applications. Dutch research and development efforts have resulted in improved system concepts, including methods for interconnecting fuel cell stacks that improve efficiency, level stack temperature, and reduce the number of heat exchangers [1].

Other developers of MCFC systems and their largest demonstration system size are as follows: Hitachi, 250 kWe system; Mitsubishi, 200 kWe system; and Ansaldo Richerche, 250 kWe system [64].

Challenges

The major challenge involved with the commercialization of the MCFC is meeting the 40,000-hour service life for the system. The issues to be solved in the MCFC technology prior to its commercialization are as follows [62, 63, 65, 66]:

- *Reduce the rate of cathode dissolution.* Dissolution of the cathode and deposition of the Ni within the electrolyte limit stack life under atmospheric pressure conditions to about 25,000 hours.
- *Reduce separator plate corrosion.* During operation, a protective oxide layer is formed in the chemically active areas of the separator plate, which impairs the hydraulic behavior, especially at the anode, by diffusion of this film into the nickel layer. For most corrosion-resistant coatings, the lifetime seems to be limited to the order of 10,000 hours.
- *Improve retention of the electrolyte.* The electrolyte is consumed in the cell by several processes, such as lithiation of the cathode, the formation of lithium chromite or lithium aluminate in the anode, transformation of aluminum oxide fibres in the matrix, and the initial corrosion of the cell hardware. Although the potential of using porous ceramic membranes to control the diffusion of electrolyte vapors in the anodic compartment has been demonstrated [67], the necessary lifetime requirements have not been demonstrated. Periodic electrolyte addition into the cells may be necessary during its service life.

- *Improve the resistance of the catalyst to poisoning* (for internal reforming only). The evaporation of the electrolyte from the cell components and condensation of it on the catalyst (coldest spot of the cell) results in the deactivation of the catalyst. A possible solution to this effect would be the utilization of the more expensive indirect internal reforming option.

Economics

A study by the Netherlands Energy Research Corporation for the Dutch Fuel Cell Corporation indicates that for 200 MWe annual production rate, stack costs of $500/kWe are feasible, including assembly, pressure vessel, and acceptance testing at the factory. However, the material costs may increase by as much as 25%, if one accounts for internal reforming. The same study estimates that precommercial units will be available at a capital cost of $2,000 to $4,000/kWe by 2000 [62]. Other estimates put the stack cost at $400/kWe and total system installation cost at $1,000/kWe [68].

In their utility-scale fuel cell report for the Japanese market, Watanabe et al. state that the future construction cost for an MCFC plant is estimated to be $2,800/kWe with the stack comprising 25% to 35% of the total system cost [69]. The market size for fuel cells is estimated as 2,000 MWe in Japan by 2010, half of which will be utility-scale plants installed by electric companies. The market share for MCFCs is estimated to be 90% with the remaining 10% of the market claimed by PAFC systems [69].

SOLID OXIDE FUEL CELLS

The solid oxide fuel cell operates in the 850°C to 1,100°C (1,550°F to 2,000°F) temperature range. At this temperature, exhaust from the SOFC can be used to power a gas turbine or a steam bottoming cycle, thus raising the overall electrical conversion efficiency to as high as 70%. It is believed that SOFCs will be commercially available during this decade in stationary combined heat and power applications in the range from 100 kWe to 7 MWe and even higher power capacities. The performance, ongoing research activities, recent commercialization efforts, and challenges associated with the development of SOFC technology are described in the following sections.

Performance and Operating Conditions

The power density of SOFCs for common operating conditions of approximately 1,000°C (1,800°F) and 3 atm is typically 0.2 to 0.5 W/cm^2.

However, Sasaki et al. report achieving a single cell power density as high as 0.9 W/cm^2 with a tubular SOFC [70].

Pressurizing the SOFC system increases the cell voltage, efficiency, and power output. Pressurizing the cell to 6 to 7 atmospheres can increase the power output by 10% to 15% over the atmospheric pressure operating conditions. Pressurization also leads to reduced capital costs due to smaller gas flow channels. The exhaust from pressurized cells can be used to drive a turbine to provide additional power. Westinghouse (currently Siemens-Westinghouse) has developed a pressurized SOFC gas turbine system [71, 72].

With high-temperature operation, one of the important issues is the time required to reach operating temperature and the response to load changes. Achenbach [73] developed an analytical model of the transient response of an SOFC to load changes. The results predicted the relaxation time based on the thermal properties, size, and configuration of the cell and the operating conditions. The rapid start-up and operating characteristics of small-diameter tubular SOFCs are discussed by Kilbride [74]. Lowering the operating temperature could lead to a more rapid start-up, reduced cost, and increased life. However, the resistance of the cell components, especially the electrolyte, increases significantly for operating temperatures lower than 630°C (1,200°F) [75]. A study by Huijsmans et al. on intermediate-temperature SOFCs indicates that the use of alternative cell materials for lower operating temperatures between 550°C and 750°C (1,000°F to 1,400°F) with ferritic steel separator plates is promising [76].

The fuel reforming in SOFC systems is another important issue. In the earlier years of development of SOFCs, the fuel was reformed externally [77]. As the technology has developed, internal reforming of the fuel stream is more common. Egussi et al. discussed the power generation characteristics of solid oxide fuel cells with internal steam reforming [78]. Internal reforming can lead to carbon formation in the anode chamber and nonuniform temperature distribution within the cell. These problems can be avoided by optimizing the internal and pre-reforming in SOFC systems based on experimental performance data [79]. Table 2-6 indicates the effects of various fuel gas species on fuel cell performance.

Materials of Construction and Manufacturing

The SOFC differs from the other fuel cells by having a solid electrolyte, allowing greater flexibility in cell configuration. SOFCs are constructed in planar, tubular, and monolithic configurations, where the

**TABLE 2-6:
Effect of Fuel Gas Constituents on
Fuel Cell Performance [22]**

Gas Species	Effect
H_2	Fuel
CO	Fuel
CH_4	Fuel (if internally reforming) or diluent
CO_2 and H_2O	Diluent
Sulfur compounds (e.g., H_2S)	Poison (> 1 ppm)

latter gets the least attention in the literature because of its relatively higher manufacturing costs. SOFCs are constructed primarily from ceramic-based materials that can withstand the high operating temperatures.

Electrolyte. The electrolyte for the SOFC should have high oxide ionic conductivity, low electronic conductivity, high chemical stability in both oxidizing and reducing atmospheres, and structural stability under the operation and fabrication conditions. Yttria-stabilized zirconia (YSZ) and tetragonal-zirconia-polycrystalline are considered as suitable electrolyte materials [80]. Current designs operating at roughly 1,000°C (1,800°F) use a YSZ electrolyte layer with a thickness of 30 to 40 μm. In order to reduce the effects of thermal stress and to facilitate the use of less exotic materials, manufacturers are exploring the development of SOFC cells that can operate at temperatures of roughly 650°C (1,200°F). At low temperatures, the resistivity of the present YSZ electrolyte layers becomes excessive. Research has focused on developing alternative electrolyte materials including very thin YSZ electrolyte layers [81], Perovskite materials [82], LaSrGaMnO electrolytes [83], as well as ceria and bismuth oxide [80].

Electrodes and Interconnect. The cathode must be designed to have sufficient porosity to transport reactant and product gases, sufficient catalytic activity to dissociate oxygen molecules, high electrical conductivity, high structural, thermal, and chemical stability, and compatibility with the electrolyte material. Several composites of lanthanum and metal oxides fulfill these requirements and detailed material studies indicate that Sr-doped lanthanum manganite ($LaMnO_3$) is a suitable cathode material. Current air electrode-supported (AES) tubular designs utilize cathodes with 30% to 40% porosity and roughly 2 mm thickness [5, 80].

The anode (fuel electrode) is not required to have catalytic activity as the chemical reactions are expected to proceed on their own at the high operating temperatures of the SOFC. Nickel in the form of an Ni-YSZ cermet is compatible with the YSZ electrolyte and is believed to be the most promising material for the anode. Current AES tubular designs utilize anodes with 20% to 40% porosity and roughly 150 μm thickness [5, 80].

An interconnect (separator) material is necessary to connect the adjacent cells electrically and prevent the fuel and oxidant gases from mixing. The requirements for the interconnect are high electrical conductivity and low ionic conductivity, high stability under the oxygen partial pressure gradient at 1,000°C (1,800°F), and thermal expansion comparable with other cell components. Ceramic materials based on $LaCrO_3$ are considered promising materials for the interconnect [80].

Manufacturing. The main manufacturing methods for the SOFC involve depositing thin films of materials on one another to form a cell structure. Typical methods for building the films are dry pressing, tape casting, tape calandering, screen printing, slurry coating, electrochemical vapor deposition (EVD), physical vapor deposition, thermal spraying, and sintering [5, 80]. Among these methods, EVD seems to be the most successful method in preparing the electrolyte, which consists of dense YSZ film on a porous substrate. Plasma spraying and sintering are likely to be the manufacturing methods of choice for the interconnection and the fuel electrode, respectively [84]. Manufacturing techniques differ with the cell geometry.

Planar SOFCs. Choy et al. describe the development of cost-effective fabrication methods of thin film electrolyte structures and LSGM ($La_{0.9}Sr_{0.1}Ga_{0.8}Mg_{0.2}O_{3-x}$) electrolytes for intermediate operating temperatures of 700°C to 800°C (1,300° to 1,500°F). Fabrication methods include electrostatic-assisted vapor deposition (EAVD) and the flame-assisted vapor deposition (FAVD). Their results show the technical viability of both EAVD and FAVD methods in the cost-effective fabrication of SOFC components and additionally that the EAVD could be used for the manufacturing and development of SOFC components operating at intermediate temperatures [85].

Drenckhahn suggests the utilization of metallic materials along with ceramic electrodes in planar high-temperature SOFCs. He states that using metallic bipolar plates fulfills the function of separating the feed gases while ensuring even distribution through the cell by having good thermal and electrical conductivities at the operating temperatures [86]. Song et al. have discussed the fabrication of planar SOFC by a composite plate process [87]. Elangovan et al. investigated the planar SOFC system technology developments by SOFCo (a McDermott Technology Inc./Ceramatec partnership). SOFCo has demonstrated stacks with less than 0.5% performance degradation per 1,000 hours for 46,000 hours of operation at 850° to 900°C (1,550°F to 1,650°F) [88].

Tubular SOFC. The main tubular SOFC developers are Siemens-Westinghouse and Mitsubishi [89]. The Mitsubishi design consists of cascaded single SOFC elements that form the SOFC tube. The Siemens-Westinghouse fuel cell consists of a tubular structure with concentric elements for the cathode, electrolyte, and anode. To simplify manufacturing and reduce cost, Siemens-Westinghouse developed an SOFC in which the structure is supported by the air electrode instead of by a separate support tube. The air electrode-supported solid oxide fuel cell was tested by Bessette and George at atmospheric and elevated pressures. The cell exhibited no noticeable decrease in performance and almost no voltage degradation for operating pressures between 1 to 10 atmospheres and for operating times up to 6,000 hours [90].

Demonstration and Commercialization Projects

The individual SOFC commercialization efforts by the major developers are described below.

Seimens-Westinghouse Corporation. The company has been developing mainly tubular SOFC power generation systems since the 1960s [91]. Working in partnership with the U.S. Department of Energy, the company is developing SOFC systems to operate on natural gas or gasified coal. The goal is to achieve 50% electrical efficiency for the fuel cell systems alone and 70% electrical efficiency for the fuel cell coupled with a gas turbine. The program goal is to commercialize the tubular SOFC by 2003 [92]. One of the major developments in conjunction with the joint project was the delivery of a 100 kWe SOFC cogeneration system in 1993 [77]. Siemens-Westinghouse focuses on a product range from 250 kWe to 7 MWe with standard 500 kWe modules. The standard power systems will include an atmospheric pressure system for cogeneration applications and a pressurized system integrated with a gas turbine (PSOFC/GT), which can provide 62% and higher efficiencies when running on natural gas [72]. The first commercially available SOFC systems will be available in 2001 [93]. The system will yield a

30% to 45% advantage in cost of electricity compared to current combustion turbine technology [71]. Two pilot plants have demonstrated a performance degradation of only 0.5% per 1000 hours of operation.

Sulzer, a Swiss power company, is working on the commercialization of an SOFC cogeneration system for residential applications. Their design is based on a planar disk configuration with integrated heat exchanger and internal steam reformer. The commercialization is expected after 2001 with a capacity range of 1 to 10 kWe.

SOFCo and its affiliated companies, Ceramatec and McDermott International Inc., are also researching systems for distributed power applications, including remote residential and small commercial systems in the 5 to 20 kWe range. Upon entering the market, the company is planning to develop 50 to 200 kWe systems. Their system design is based on the use of a SOFC stack, an advanced heat exchanger, a steam generator, a desulfurizer, a fuel processor, and a start-up burner. The latest prototype is a 2 kWe system.

Other companies that have SOFC systems in the laboratory stage of development include Allied Signal, Ztek, Rolls Royce, and Mitsubishi [1].

Economics

The goal of Siemens-Westinghouse is to commercialize the PSOFC/GT power systems with fully installed costs of under $1,300/kWe at a market entry level for 1 to 3 MWe power output sizes. The conceptual design for a commercial manufacturing facility has been performed and the facility and equipment cost for the first 100 MWe/year cell production line was estimated to be $80 million. [93].

Itoh et al. have developed a production cost estimation for Japanese companies manufacturing SOFCs using electrochemical vapor deposition (EVD) and plasma spraying (PS) methods. Their estimates are based on production rates of 2,000 kWe/month. The planar SOFC cost is estimated as $1,100/kWe, while the tubular SOFC cost is estimated as $5,900/kWe with EVD and $2,800/kWe with plasma spraying [94].

A study by Ippommatsu et al. yielded more promising results [95]. Assuming mass production on the order of 1 million cells per plant per year (each cell has 400 cm^2 active cell area), Ippommatsu et al. estimated the manufacturing costs for the four major techniques (EVD, EVD + laser ablation, plasma spray, and sintering). The costs of the production processes for a cell varied from $54 to $67. Assuming a power output of 0.3 W/cm^2, the manufacturing cost for a cell would be $450 to $780/kWe. These values make the SOFC competitive with existing technologies. If the power density can be increased to 0.5-0.7 W/cm^2, the SOFC is believed to be more cost-effective.

FUEL CELL TECHNOLOGY FORECAST

In the current literature, there is considerable uncertainty regarding the time frame for introduction of fuel cell technology, the anticipated cost of the technology, and the expected life of a fuel cell stack. To help clarify these issues, an opinion survey was conducted among manufacturers and researchers with expertise in fuel cell technology. The survey was designed to assess the time to limited and widespread commercial availability, the expected first cost in the near and long term, anticipated annual maintenance costs, and the expected life of the stack. In addition, the survey solicited opinions regarding the applicability of the various technologies to cogeneration applications in buildings, critical maintenance issues, technological barriers to the development of fuel cell cogeneration, and the availability of incentives for adoption of fuel cell technology. Fourteen responses to the survey were received with applicants expressing expertise in the following areas:

PEMFC	11
PAFC	7
MCFC	6
SOFC	6
Fuel processing	5
Electrical systems	4

Summaries of responses to the survey are presented in Figures 2-1 and 2-2.

Figure 2-1a summarizes the opinions of the respondents regarding the applicable size range for each of the fuel cell technologies. For applications involving less that 100 kWe, such as residential and small commercial buildings, PEMFC technology appears to be the most promising. For systems in the range of 100 kWe to 1 MWe, such as small to mid-size commercial buildings, PAFC, MCFC, and SOFC are all considered promising. For large building and utility applications requiring more than 1 MW, MCFC and SOFC systems are likely to be attractive.

Figures 2-1b and 2-1c present the respondents' forecasts for the dates of initial and mass market commercialization for each technology. Most respondents foresee commercialization in limited markets for each of the technologies prior to 2006. Most respondents acknowledge the current availability of PAFC systems and expect PEMFC systems to be the next technology to reach the market followed by MCFC and SOFC

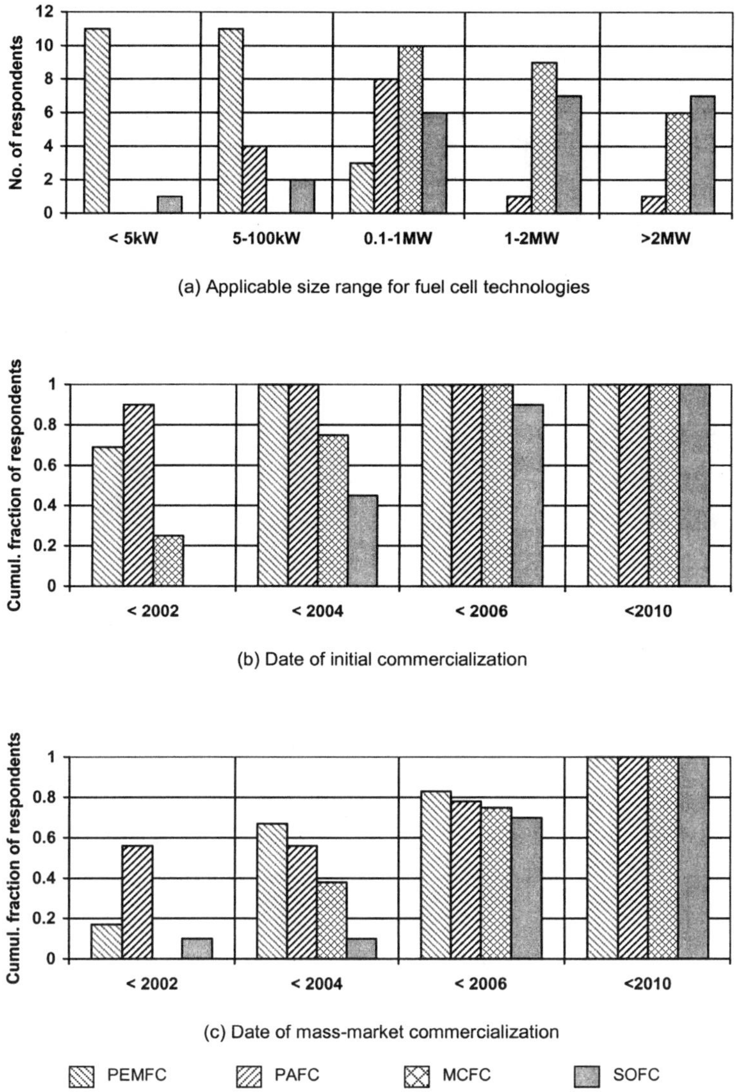

Figure 2-1 Forecast of applicability and availability of fuel cell system technology.

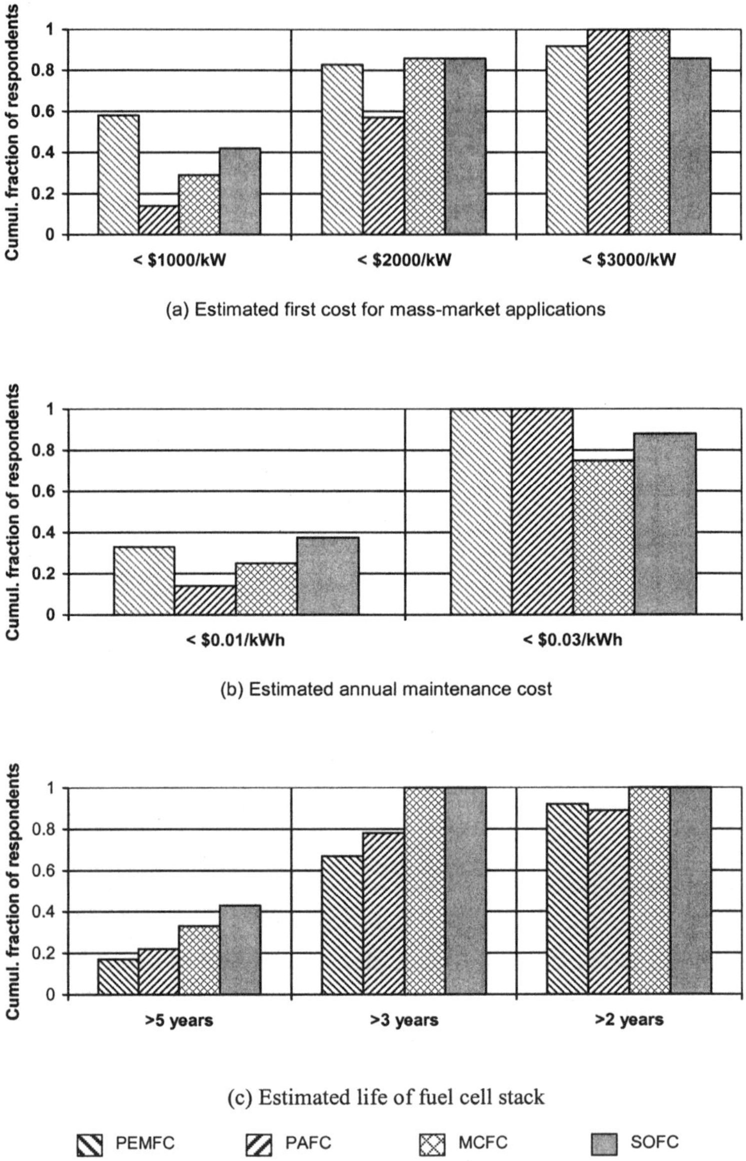

Figure 2-2 Forecast of economic parameters for fuel cell systems.

systems. Acceptance in wider markets will likely take longer. However, respondents forecast that all four technologies will be available in mass markets by 2010, and 60% to 80% of respondents expect mass market commercialization of each technology prior to 2006.

Figure 2-2a reflects the opinions of the respondents regarding the first cost of each of the fuel cell system technologies. Better than 80% of respondents predict that first costs for PEMFC, MCFC, and SOFC systems will be less than \$2,000/kWe. Fewer respondents foresee PAFC technology achieving this goal. Respondents were most optimistic about the prospects of PEMFC and SOFC technology achieving initial costs less than \$1,000/kWe.

The opinions regarding annual maintenance cost were somewhat discouraging. As reflected in Figure 2-2b, less than 40% of respondents believe that maintenance costs (including a sinking fund for stack replacement) will be less than \$0.01/kWh. Most respondents foresee fuel cell stack lives exceeding three years for PAFC, MCFC, and SOFC systems, but most respondents do not believe that stacks will last longer than five years. PEMFC stacks will likely have the shortest life, and only 67% of respondents foresee PEMFC stack lives exceeding three years.

In general, respondents to this survey seem to believe that over the next five years, fuel cell systems are likely to be introduced for a variety of applications with PEMFC systems applicable in the smaller size ranges and MCFC and SOFC systems attractive in the larger size ranges. Respondents seem optimistic that cost targets of less than \$2,000/kWe are attainable, with many respondents expecting PEMFC and SOFC systems to eventually be available for less than \$1,000/kW. Fuel cell system maintenance costs are currently expected to be somewhat higher than costs for conventional on-site generating technologies. Fuel cell stack life, expected to range from three to five years, is likely to be adequate. However, an improvement in stack life and a corresponding reduction in the annual stack replacement allowance (included in the maintenance cost) would likely improve the economic viability of fuel cell systems.

Chapter 3

ASSESSMENT OF FUEL CELL SYSTEMS

As discussed in Chapter 1, there are four basic fuel cell *stack* technologies that use hydrogen-rich fuel to produce direct current and thermal energy. Fuel cell *systems* include not only the fuel cell *stack* but also

- a *fuel processor* to allow operation with available hydrocarbon fuels,
- a *power conditioner* to regulate the output power of the cell and where necessary convert it to alternating current,
- an *air management system* to deliver air at the proper temperature, pressure, and humidity,
- a *thermal management system* to remove heat from the stack and to transfer heat among the various system components and, in some cases,
- a *water management system* to ensure that water is available for fuel processing and reactant humidification.

Figure 3-1 presents a schematic of a fuel cell system. This chapter describes each of the major subsystems (except for the stack itself, which was discussed in Chapters 1 and 2) that are required for fuel cell system operation. For each of the major stack technologies, a representative design is presented to illustrate how the subsystems work together. Finally, general characteristics of fuel cell systems are reviewed and compared to characteristics of other on-site generating technologies.

SUBSYSTEMS

Fuel Cell Stacks

The four major types of fuel cell stacks—PEMFC, PAFC, MCFC, and SOFC—were described in Chapter 1. The low-temperature stacks,

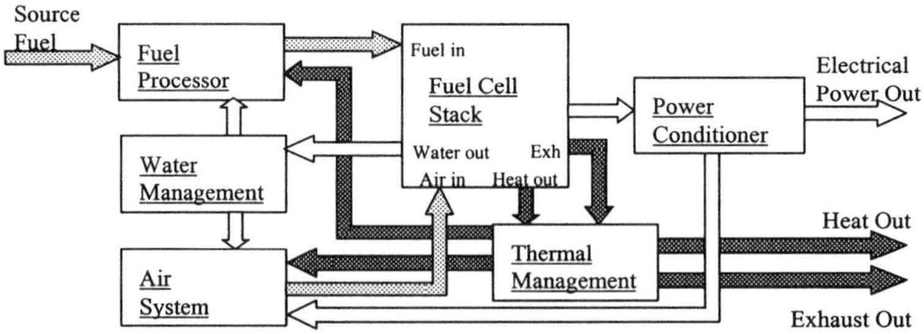

Figure 3-1 Fuel cell system schematic.

PEMFC and PAFC, require more extensive fuel processing and yield thermal energy at a lower temperature. The high-temperature stacks, MCFC and SOFC, are more flexible in their fuel requirements and yield thermal energy at a higher temperature. All four types have similar requirements with respect to power conditioning.

Fuel Processing

Low-temperature fuel cells, such as PEMFCs and PAFCs, require hydrogen for operation. Higher temperature cells, including MCFCs and SOFCs, can use hydrogen and carbon monoxide as fuel and may even be able to reform a simple hydrocarbon, such as methane, to a usable fuel within the cell. To supply fuel to the low-temperature cells and to permit the use of fuels other than H_2, CO, or methane in high-temperature cells, fuel reformers are required. In addition, virtually all fuels require some type of "clean-up" operation to remove contaminants such as compounds of sulfur. The combination of clean-up and reforming is referred to as fuel processing. The following sections describe the major types of reformers and the processes available for gas clean-up.

Steam Reforming. Steam reforming is a well-established industrial technology for producing H_2 and CO from hydrocarbon fuels. Steam reforming involves mixing steam with the hydrocarbon fuel at high temperatures to form H_2 and CO according to the reaction

$$C_nH_m + n\mathrm{H_2O} \leftrightarrow (n + m/2)\mathrm{H_2} + n\mathrm{CO}. \tag{3-1}$$

Equation 3-1 is developed for a generic hydrocarbon fuel. The specific steam reforming reaction for methane was given in Equation 1-41. Steam reforming is most applicable to low carbon fuels, including methane, butane, and propane, but even naptha has been reformed with this technique [22]. The reforming reaction occurs over a nickel-based reforming catalyst with high temperature and low pressure favoring the products. At typical temperatures of 800ºC to 900ºC (1,500ºF to 1,650ºF), a steam-to-carbon feed ratio of 3.5, and moderate pressures, approximately 98% of methane is converted to H_2 and CO [4]. The product gas can be used directly in MCFCs or SOFCs, but for PEMFCs and PAFCs, further processing is required to eliminate residual CO.

The steam reforming reaction is highly endothermic. In addition, heat is required to produce steam and to heat the methane to operating temperature. Proper integration of the reformer heat requirements with heat available from the reformer exhaust and from the fuel cell is essential for efficient operation. Some reformers use a regenerative design such as that illustrated in Figure 3-2.

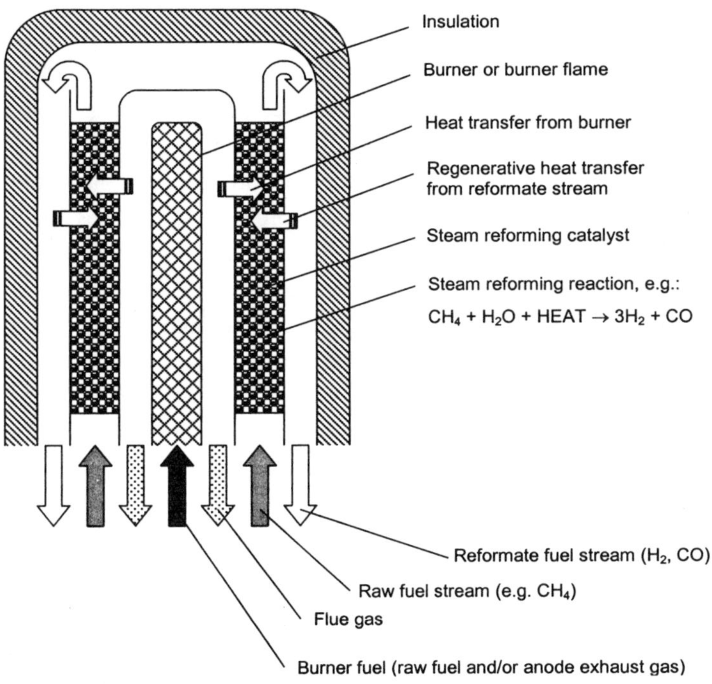

Figure 3-2 Regenerative steam reformer (adapted from [5]).

Partial Oxidation. Partial oxidation is a process that is used to reform heavy hydrocarbons at very high temperatures. The basic conversion reaction is the same as the steam reforming reaction, Equation 3-1. But in partial oxidation, the heat for the reaction is supplied by combusting a fraction x of the feedstock C_nH_m with oxygen or air in a refractory-lined reactor [4]:

$$xC_nH_m + (nx + mx/4)O_2 + 3.76(nx + mx/4)N_2 \rightarrow (mx/2)H_2O + nxCO_2 + 3.76(nx + mx/4)N_2. \tag{3-2}$$

The net reaction is

$$C_nH_m + (nx + mx/4)O_2 + 3.76(nx + mx/4)N_2 + ((1-x)n - mx/2)H_2O \rightarrow (n + m/2)(1-x)H_2 + n(1-x)CO + nxCO_2 + 3.76(nx + mx/4)N_2. \tag{3-3}$$

Providing the heat of reaction by combustion within the reactor instead of by heat transfer through the tube walls permits much higher reaction temperatures than are possible in conventional steam reforming. At typical operating temperatures of 1,300ºC to 1,500ºC (2,400ºF to 2,800ºF), the partial oxidation reaction can proceed without catalyst. Heat from the reaction can be recovered to preheat the feed gas and produce steam for the reaction. The product gas can be used directly in MCFCs or SOFCs, but for PEMFCs and PAFCs, further processing is required to eliminate CO. The disadvantage to partial oxidation is that if air is used for oxidation, then the resulting fuel stream will contain carbon dioxide, nitrogen, and possibly ammonia. In addition, since some of the fuel is combusted instead of reformed, the hydrogen yield with partial oxidation is lower than with steam reforming.

Autothermal Reforming. Autothermal reforming uses partial oxidation of the fuel stream in the presence of a reforming catalyst. For a particular hydrocarbon feed, autothermal reforming will operate at a lower temperature than partial oxidation but at a higher temperature than steam reforming. Autothermal reformers utilize a platinum combustion catalyst and a conventional nickel steam reforming catalyst. Like partial oxidation and steam reforming, autothermal reforming produces a fuel that requires further processing for use in PAFCs or PEMFCs.

Shift Conversion. Shift converters reduce the CO content and increase the H_2 content of the fuel stream through the water gas shift reaction,

$$CO + H_2O \leftrightarrow CO_2 + H_2. \tag{3-4}$$

This reaction is exothermic with lower temperatures favoring the products. Shift conversion is often accomplished in two temperature steps with catalysts selected for each step based on the operating temperature. High-temperature shift converters typically operate at temperatures of 330°C to 530°C (630°F to 990°F) and use iron and chromium oxide as the catalysts. Gas leaving the high-temperature shift converter contains 1% to 2% CO by volume. This gas is cooled and then fed to a low-temperature shift converter to further reduce the CO content. The low-temperature shift catalyst is commonly copper and zinc oxide supported on alumina and is active in the range of 200°C to 250°C (390°F to 480°F) [4]. The CO content of the gas leaving the low-temperature converter can be as low as 0.5%. This level is adequate for all cells except PEMFCs, which require further processing to reduce CO to the ppm range.

Gas Clean-Up. Gas clean-up refers to a variety of processes that can occur upstream or downstream of the fuel reformers and shift converters. Catalysts used in steam and autothermal reformers, shift converters, and in the fuel cell itself are generally sensitive to compounds of sulfur. Many commercial fuels contain sulfur naturally or have sulfur added as an odorant to facilitate leak detection. Thus, one of the first gas cleaning operations is *desulfurization*. This can be accomplished by activated carbon adsorbers and/or by zinc oxide sulfur polishers operating at a temperature of 350°C to 400°C (660°F to 700°F). Certain sulfur compounds may require reaction with hydrogen to convert them to H_2S, which is easily removed by zinc oxide [22].

PEMFCs require CO concentrations as low as 10 ppm. To achieve these low levels, *selective catalytic oxidation* may be employed downstream of the shift converters. With selective oxidation, oxygen is supplied to the reformate in the presence of a selective catalyst. The catalyst is designed to favor the oxidation of CO over that of H_2, thus converting CO to CO_2, which is a diluent in the fuel stream but not a poison to the PEMFC. As an alternative, the CO concentration can be reduced by *pressure swing adsorption*, *methanation*, or by *membrane separation*.

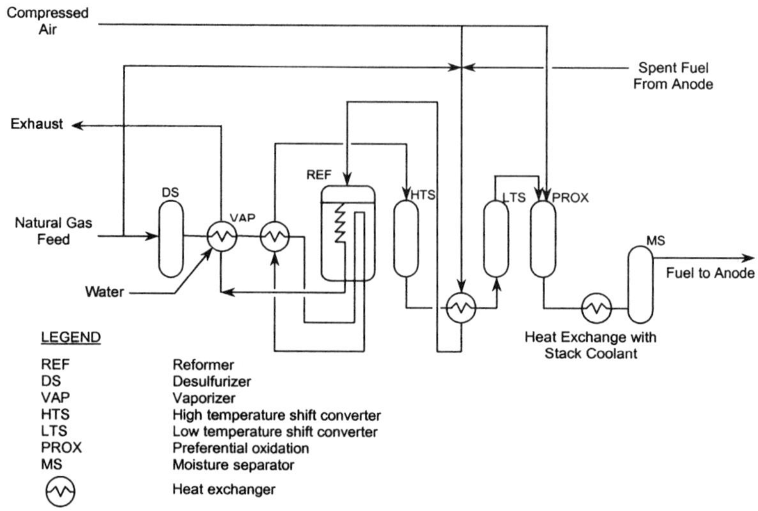

Figure 3-3 Fuel processing system for PEMFC—natural gas feedstock.

Figure 3-3 illustrates some of the fuel processing operations as applied to the reformation of natural gas (primarily methane) for use in a PEMFC. The natural gas is first fed to a desulfurizer to remove sulfur to protect the fuel cell and reforming catalysts. The gas is then combined with water and reacted over a catalyst in a steam reformer. The effluent from the steam reformer enters a high-temperature shift converter where CO is exchanged for H_2 in accordance with the water gas shift equation, Equation 3-4. A second low-temperature shift converter further reduces the concentration of CO to roughly 0.5%. Finally, a selective catalytic oxidizer oxidizes the remaining CO to CO_2. The final feed stream to the fuel cell consists of a mixture of H_2, CO_2, H_2O, and trace amounts of CO. The feed stream is cooled to the cell-operating temperature and condensate is removed prior to introducing the fuel into the anode manifold. Unreacted fuel from the anode is combined with natural gas and air and combusted to provide heat for the reforming process.

Clearly, the fuel processor is a complex system consisting of a number of different components. Consequently, it is difficult to scale economically to small sizes. In addition, due to heat and mass storage effects in the reactors, start-up times tend to be long and the response to load changes tends to be slow. The development of compact, efficient, and economical fuel processors that respond quickly to load changes is currently an active area of research.

Power Conditioning

The power conditioning system modifies the fuel cell electrical output to meet the needs of the application, provides power for fuel cell auxiliary systems, and interfaces with the fuel cell control system. The features incorporated into the power conditioning system depend upon the type of application. There are four basic types of fuel cell power applications:

- *Grid independent* applications in which the fuel cell is the sole provider of power to the load. Examples include remote and portable power applications.
- *Back-up power* applications in which two power sources are required in order to improve reliability. In this case, the utility grid may be the primary power source with the fuel cell providing power only in the event of a failure of the grid. Alternately, the fuel cell may provide power with the utility acting as a backup. In either case, the fuel cell and the grid do not provide power simultaneously.
- *Parallel power* applications in which the fuel cell and the utility power source operate simultaneously to meet the power requirements of the application.
- *Utility interconnect* applications in which the fuel cell not only provides power to the application but also can supply power into the utility grid.

In all four types of applications, the power conditioning system

- regulates the voltage provided by the fuel cell system,
- transforms DC power from the fuel cell into AC power for the application,
- provides reactive power to match the power factor of the application,
- provides power to fuel cell auxiliary devices, and
- interfaces with the electrical load and the fuel cell system controls.

The basic features of voltage regulation, conversion to AC power, and provision of reactive power are accomplished very efficiently by modern solid state power electronics. At full load, the power conditioning system typically supplies approximately 95% of its input power as real power to the load or to auxiliary devices.

In addition to the basic functions listed above, backup power applications require that the power conditioning system monitor the utility system and initiate the supply of fuel cell power in the event of a loss of utility power. Parallel power applications also require that the power conditioner synchronize with the voltage and current waveforms that are provided by the utility grid. The parallel power system must also include interlock relays that prevent the fuel cell system from feeding power back into the utility grid. Systems that are fully interconnected with the utility grid such that they can supply power into the grid as well as to the application require additional features to meter the flow of power into the grid, to safeguard the utility grid, and possibly to permit utility control of the fuel cell system.

The interconnection of distributed power systems such as fuel cells with the utility grid is currently an important area of development. The present utility grid is essentially designed to be a one-way conduit for power. Features providing safety and maintainability for the present system are often inadequate for systems in which power can travel in both directions. Systems that are intended for parallel operation and for full utility interconnection must be designed to meet governing codes *and* the requirements of the local electric utility. In order to facilitate the development of standardized interconnects, the IEEE is currently developing Standard P1547, Standard for Interconnecting Distributed Resources with Electric Power Systems. This standard will address requirements "relevant to the performance, operation, testing, safety considerations, and maintenance of the interconnection" [96].

Air Management

Air is required for the fuel cell stack and for the fuel processor. Air is generally supplied by a compressor at pressures of 2 to 10 atmospheres depending on the fuel cell technology. The flow rate of air is determined by the reaction rate of the oxygen at the cathode. As the air travels through the cathode, oxygen is depleted from the air. In order to provide an adequate concentration of oxygen throughout the cathode, air is generally supplied at a rate that exceeds the rate required by the reaction. The stoichiometric ratio is the ratio of air actually supplied to the minimum amount of air required by the reaction. Stoichiometric ratios of 2 are commonly employed. The pressure of the air supplied by the compressor must be at least high enough to overcome the pressure losses through the air manifolds and the cathode flow channels. In addition, the cathode compartment of the fuel cell may be operated at a pressure above atmospheric in order to improve stack performance and/or to facilitate humidification of the airstream.

The air system has an important effect on the fuel cell system performance. In general, as the air pressure increases, activation and concentration polarization losses are reduced, yielding higher stack power density and higher efficiency. However, the power required to operate the compressor also increases with pressure. If the compressor power is provided by the fuel cell stack, the net power from the fuel cell *system* is reduced. Thus, the best system performance occurs at an optimum operating pressure that represents a balance between increased stack performance and increased compressor power requirements. In order to reduce the parasitic effect of the air compressor at all load conditions, part of the compressor power can be supplied by a turbine or other expansion device located in the stack or reformer exhaust stream.

When the fuel cell system is operating at part load, the performance of the air compressor can become particularly important. If the air compressor power remains constant at part load, the fraction of the gross power that goes to run the compressor will increase and the fuel cell system efficiency will decrease. Thus, variable displacement or variable speed compressors are often employed so that the airflow rate and compressor power can be reduced at part load.

In addition to the air compressor, the air supply system will include heat exchangers and humidifiers (for the PEMFC) to ensure that the temperature and humidity of the airstream are compatible with the stack.

Thermal Management

The thermal management system consists of the network of heat exchangers, fans, pumps, and compressors that are required to heat and cool the various process streams entering and leaving the fuel cell and fuel processor. The configuration of the thermal management system varies with the type of fuel cell and examples of several system configurations are illustrated in later sections. However, all systems must provide for stack cooling, heat recovery for cogeneration, and reactant preheating or precooling.

The amount of heat released during the fuel cell reaction is comparable to the amount of electricity that is provided. This heat must be removed from the fuel cell stack. A portion of it can be recovered to meet the thermal needs of a cogeneration application. In PEMFC and PAFC stacks, stack cooling is accomplished by circulating a heat transfer fluid through cooling channels in the stack. Heat transferred to the coolant can be recovered and used to meet the thermal needs of the building application. In MCFC and SOFC applications, cell cooling is provided by the anode and cathode gas streams, which leave the stack at a higher temperature than they enter. In these high-temperature stacks, heat transfer from the cell reaction may also be used to supply energy to the endothermic fuel reforming reaction occurring within the cell stack.

In the MCFC stack, the gas leaving the anode may be cooled prior to being mixed with incoming air and reintroduced in the cathode. In these higher temperature stacks, heat recovery for cogeneration is typically accomplished from the exhaust gas stream instead of from a stack coolant fluid.

Fuel and air streams must be conditioned to the proper temperature before entering the cell stack. If the fuel is supplied from a fuel processor, it will need to be cooled before entering a PEMFC or PAFC stack. The heat transferred from the fuel stream can be used to preheat the feed gas to the fuel processor or to produce steam for fuel processing. In the MCFC and SOFC, the fuel stream is preheated with energy transferred from the cell exhaust.

On the airside, hot air leaving the compressor may need to be cooled before it enters the PEMFC stack. This can sometimes be accomplished by spraying water into the air, thus accomplishing humidification as well. In the PAFC stack, the hot air leaving the compressor is supplied directly to the stack. In the MCFC stack, the air entering the stack is preheated by mixing it with hot gas from the anode exhaust stream, thus accomplishing air heating and CO_2 recycling. In the SOFC stack, air is preheated with energy transferred from the stack exhaust.

Water Management

Water is required for fuel processing in all fuel cell systems and is also required for humidification of the reactant gases in the PEMFC. Since water is produced by the fuel cell reaction, it can often be condensed from the exhaust stream and reused. If the condensed water does not meet the water requirements, additional water from a source of high purity must be provided.

REPRESENTATIVE FUEL CELL SYSTEM DESIGNS

To illustrate how the various subsystems of the fuel cell system can be integrated, the following sections describe typical systems based on each of the four fuel cell technologies. These systems are only examples, and a variety of different configurations are possible for each cell technology depending on the fuel processor, the system application, and the system design choices.

Proton Exchange Membrane Fuel Cell (PEMFC)

An example of a natural gas-fueled PEMFC system is illustrated in Figure 3-4. The fuel processor consists of a sulfur polisher, a steam reformer, high- and low-temperature shift converters, a partial oxidation

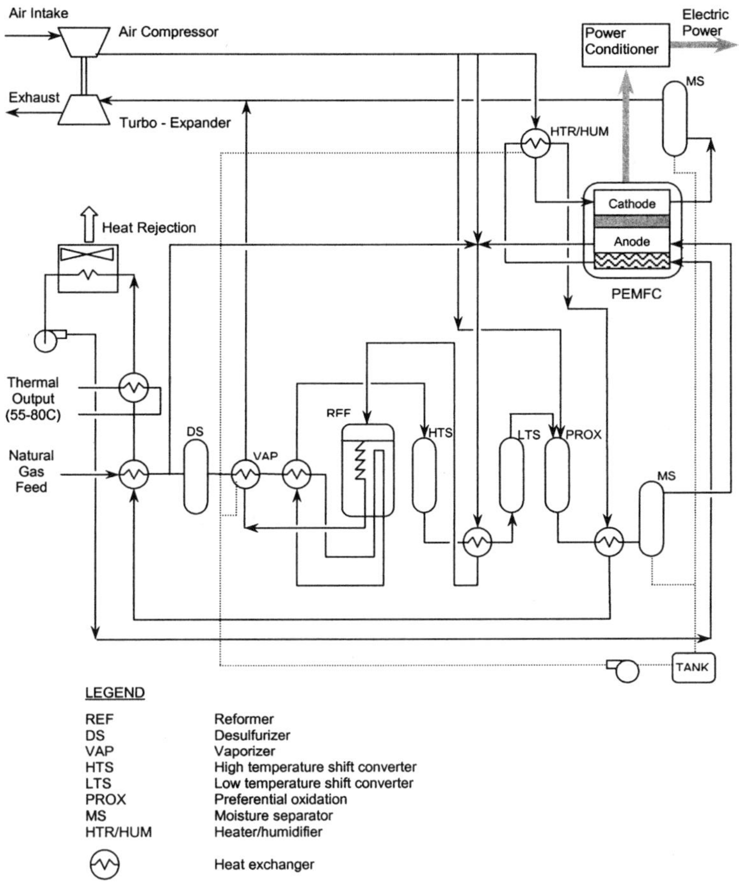

Figure 3-4 Proton exchange membrane fuel cell system.

reactor, and heat exchangers for thermal management. The air compressor that supplies air to the system is driven by a turbo-expander that recovers energy from the pressurized exhaust stream.

To understand the process in more detail, we begin with the natural gas, which is first preheated and desulfurized. Water is then evaporated into the fuel in the vaporizer and the mixture is then heated with energy recovered from the reformer product stream. The hot fuel/steam mixture then enters the steam reformer where it is heated over a catalyst to yield a stream rich in H_2 and CO. The reformer product stream passes through the feed gas heat exchanger and then goes to the high-temperature shift converter that reduces the CO content and yields additional H_2. Effluent from the high-temperature shift converter passes through a heat exchanger and supplies heat to the fuel/air mixture that is used to fire the reformer. The cooled fuel gas from the high-temperature shift converter then goes to the low-temperature shift converter where the CO content is reduced to less than 1%. The fuel then goes to the preferential oxidation reactor where CO is oxidized to CO_2 over a selective catalyst, yielding a fuel stream that has less than 10 ppm CO. The fuel is then cooled, demisted, and supplied to the anode manifold of the fuel cell stack. Depleted fuel leaving the anode is mixed with air and natural gas and combusted in the reformer. Exhaust from the reformer supplies heat to vaporize water into the fuel stream and is then expanded through the turbo-expander before being exhausted from the system.

In the air system, air is supplied by a compressor that is driven by the turbo-expander. A portion of the air is used for combustion in the reformer and to oxidize CO in the preferential oxidation reactor. The rest of the air is heated and humidified and then supplied to the cathode manifold of the fuel cell stack. Hot moist air leaving the cathode is demisted to remove water and then expanded through the turbo-expander before being exhausted from the system.

The thermal management system consists of a circulating coolant loop and associated heat exchangers. Coolant is pumped through the fuel cell stack. Hot coolant leaving the stack is used to provide energy to the air heater/humidifier. The coolant is then circulated through a heat exchanger in the fuel supply stream where it is used to cool the fuel stream to the cell operating temperature. The coolant is then used to preheat the natural gas feed stream and to supply thermal energy for purposes such as space heating or water heating. Energy that cannot be put to a useful purpose is rejected through a fluid cooler. The coolant is then returned to the cell stack.

Water management is also critical for system operation. Water is used in the fuel processor and to humidify the air supplied to the cathode. Water is available from the moisture separators in the cathode exhaust stream and the anode inlet stream. Any shortage of water must be supplied from an external source.

PEMFCs are currently under development by a number of competing manufacturers. Details of the system designs are typically not available due to proprietary concerns. A pressurized PEMFC system, similar to the one illustrated in Figure 3-4 operating on reformed natural gas, would typically operate at a pressure of 3 atm (45 psia), a power density of 0.15 to 0.25 W/cm^2, and a fuel utilization of 75% to 85% [22]. The resulting electrical conversion efficiency would be roughly 40% (based on lower heating value) with an additional 40% of the input energy available as thermal energy at a temperature of 55°C to 80°C (130°F to 170°F) [34].

Phosphoric Acid Fuel Cell (PAFC)

An example of a natural gas-fueled PAFC system [22] is illustrated in Figure 3-5. The fuel processor consists of a hydrolzer, a sulfur polisher, a steam reformer, high- and low-temperature shift converters, and heat exchangers for thermal management. The system that is illustrated operates at 8 atm of pressure. A two-stage compressor with intercooling compresses the air. The compressor is driven by a turbo-expander that recovers energy from the pressurized exhaust stream.

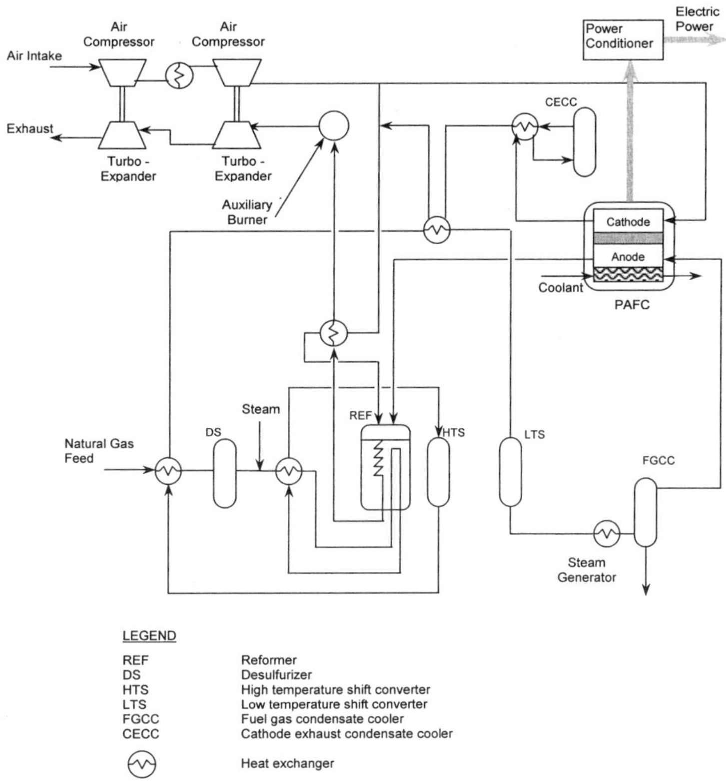

Figure 3-5 Phosphoric acid fuel cell system (adapted from [22]).

Following the fuel through the system, natural gas is first hydrolyzed to convert certain sulfur compounds to H_2S, which is then removed along with the remaining sulfur in a sulfur polisher. Steam is then added to the fuel and the mixture is heated with energy recovered from the reformer effluent. The hot fuel gas then enters the steam reformer where it is heated over a catalyst to yield a stream rich in H_2 and CO. The reformer effluent passes through the feed gas heat exchanger and then goes to the high-temperature shift converter that reduces the CO content and yields additional H_2. Effluent from the high-temperature shift converter passes through heat exchangers where it is used to preheat natural gas prior to desulfurization and to preheat the cathode exhaust prior to its use in the reformer. The cooled fuel gas then goes to the low-temperature shift converter where the CO content is reduced to 0.7%. The fuel is then cooled, demisted, and supplied to the anode manifold of the fuel cell stack. Spent fuel leaving the anode is combusted in the reformer and the auxiliary burner before it is expanded through the turbo-expanders and exhausted from the system.

In the air system, air is compressed in the two-stage compressor. A fraction of the air is used in the reforming process. The remainder is supplied to the cathode manifold of the fuel cell stack. Hot air leaving the cathode is cooled and demisted to remove water. It is then heated by energy recovered from the cathode exit air, the high-temperature shift converter exhaust, and the reformer effluent before being used to oxidize fuel in the reformer.

The fuel cell module itself consists of twelve 1 MWe stacks operating at the following conditions:

Cell voltage:	0.76 V
Current density:	320 mA/cm^2
Temperature:	207°C (405°F)
Fuel utilization:	86.2%

The gross power output from the stack is 13.25 MWe. The inverter has an efficiency of 97% (0.4 MWe loss). Auxiliary components require 0.54 MWe of power, yielding a net power of 12.3 MWe. Natural gas is supplied to the system at a rate of 25.4 MW (86.7×10^6 Btu/h) based on LHV. Based on this value, the overall system has an efficiency of 48.4% (corresponding to a heat rate of 7,050 Btu/kWh). Thermal energy from the stack coolant water that is not used in the reforming process is used to produce 2.1 MWt (7.1×10^6 Btu/h) of 12.2 atm (165 psig) saturated steam.

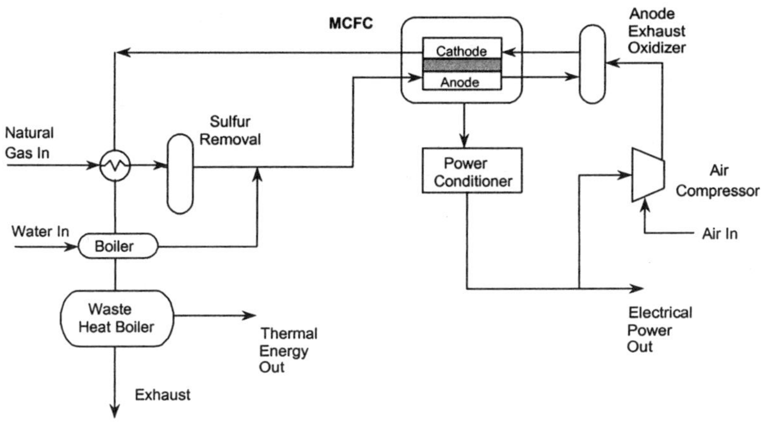

Figure 3-6 Molten carbonate fuel cell system.

Molten Carbonate Fuel Cell (MCFC)

An example of a natural gas-fueled direct internal reforming MCFC system is illustrated in Figure 3-6. The example system is based on a 300 kWe system developed by Fuel Cell Energy Systems [97]. The fuel stream is preheated and then desulfurized using activated carbon. Steam is injected into the clean fuel stream and the mixture is introduced into the fuel cell anode. In the anode, the steam reforming reaction coupled with the water gas shift reaction yields H_2 that feeds the fuel cell reaction. Depleted fuel consisting of CO_2, H_2O, and unreacted H_2 leaves the anode and is further oxidized to remove the remaining H_2 in the anode exhaust oxidizer. The anode exhaust, which is rich in CO_2, is mixed with air at atmospheric pressure and supplied to the cathode. The cathode effluent is passed through a heat exchanger where heat is transferred to the incoming fuel stream and then through a steam boiler that produces steam for the internal reformation process. Finally, the exhaust stream can be used in a heat recovery boiler to produce steam for other applications.

The fuel cell stack consists of 350 cells operating at the following conditions:

Cell voltage:	0.7 V
Current density:	200 mA/cm^2 (estimated)
Temperature:	650°C (1,200°F)
Fuel utilization:	80% (estimated)

The gross power output from the stack is 350 kWe. The inverter has an efficiency of 93% (25 kWe loss). Auxiliary components require 25 kWe of power (estimated), yielding a net power of 300 kWe. The overall system has an efficiency of approximately 55% (corresponding to a heat rate of 6,200 Btu/kWh). Thermal energy from the exhaust stream can be used to produce 140 kWt (480,000 Btu/h) of useful thermal energy for a cogeneration efficiency of approximately 80%.

Solid Oxide Fuel Cell (SOFC)

An example of a natural gas-fueled pressurized SOFC system is illustrated in Figure 3-7. The fuel stream is desulfurized prior to entering the stack. All reforming and shift reactions occur within the stack itself. Hydrogen, and possibly CO, are oxidized inside the fuel cells. Unreacted fuel and oxidant streams are mixed and combusted prior to leaving the stack, thus providing energy to preheat the incoming air stream. Incoming air is compressed to the system operating pressure of 3.5 atm and preheated before entering the stack. The effluent from the fuel cell stack is expanded through a turbine that provides power to drive the air compressor and a generator that supplies electricity. The power from the generator and the fuel cell are regulated and inverted to

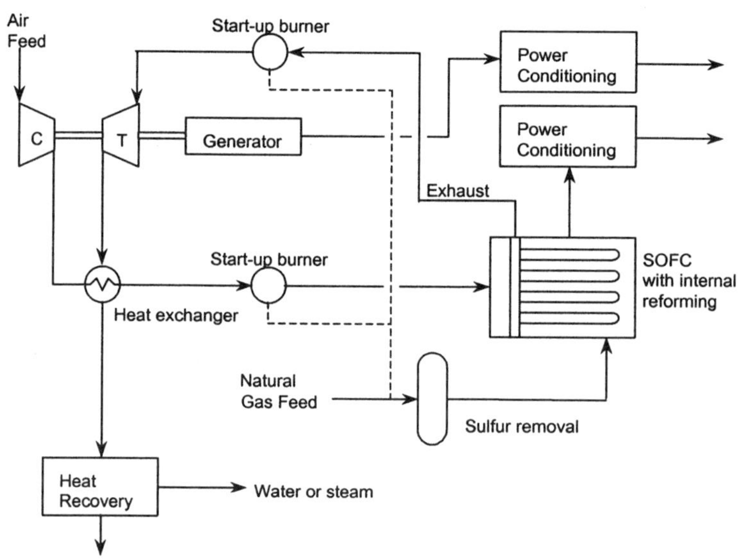

Figure 3-7 Pressurized solid oxide fuel cell system with gas turbine.

provide AC power. The exhaust gases leaving the turbine are used to preheat the air prior to entering the fuel cell stack. After leaving the heat exchanger, the exhaust gases are at a temperature of approximately 200°C (390°F) and could be used with a heat recovery boiler to provide energy for producing hot water or steam. A separate cooling system for the stack is not required since the thermal energy from the fuel cell reaction is used internally to provide energy for the endothermic reforming reaction and for heating inlet fuel and air.

The fuel cell stack consists of 1152 tubes operating at the following conditions:

Cell voltage: 0.67

Current density: 200 mA/cm^2

Temperature: 1,000°C (1,800°F)

Fuel utilization: 85%

The power output from the stack is 180 kWe. The power output from the turbine is 40 kWe. Thus, the gross power output from the system is 220 kWe. The inverter has an efficiency of 92% (18 kWe loss). The overall system has an efficiency of 60% (corresponding to a heat rate of 5,700 Btu/kWh).

GENERAL FUEL CELL SYSTEM CHARACTERISTICS

Efficiency

The overall electrical conversion efficiency of a fuel cell system is defined as the electrical power out divided by the chemical energy into the system and, as shown in Equations 1-29 and 1-30, is equal to the product of the ideal, voltage, current, fuel, system, and fuel processor efficiencies and the fuel processor mass flow ratio:

$$\eta_E \equiv \frac{\dot{W}_s}{\dot{m}_f \cdot HV_f} = \eta_{ideal}\eta_V\eta_I\eta_U\eta_S\eta_{FP}\varepsilon_{FP} \qquad (3\text{-}5)$$

The electrical conversion efficiency ranges from 35% to 55% depending on the fuel cell technology, fuel source, system design, and operating conditions.

The ideal, voltage, current, and fuel utilization efficiencies in Equation 3-5 are reflected in the fuel cell stack efficiency. The stack efficiency for various fuel cell technologies was discussed in Chapter 1. The system efficiency reflects the fuel processor efficiency, the fraction of the power that goes to operate auxiliary equipment, the efficiency of the power conditioner, and the power produced from the stack exhaust stream. Auxiliary devices, including fans, compressors, pumps, and controls, usually require roughly 5% to 10% of the gross system power at full load depending on the operating pressure and system design. The inverter efficiency, defined as the ratio of the electrical power out to the electrical power in, is typically in the range of 94% to 97%. The relative significance of the auxiliary systems can increase significantly at partial load.

As noted in Equation 1-28, the fuel processor efficiency is typically defined as

$$\eta_{FP} \equiv \frac{HV_a}{HV_f} \tag{3-6}$$

where HV_a is the heating value (HHV for PEMFC—LHV for others) of the fuel supplied to the anode and HV_f is the heating value of the fuel entering the processor. The denominator of this definition accounts only for the chemical energy entering the reforming process—not for heat transfer to the reformer. When the stack is operating at temperatures high enough to permit heat transfer to the reforming process, the reformer can actually yield an anode fuel stream that has a higher heating value than the incoming fuel stream. Thus, particularly in a high-temperature fuel cell system, the reformer efficiency can actually be greater than 100%. Little conducted a study of steam, partial oxidation, and autothermal reformers integrated into PEMFC and PAFC systems. For source fuels, including ethane, methane, and methanol, fuel processor efficiencies were found to range from 70% to 110% [2].

Figure 3-8 provides an example of representative energy flows for a PAFC fuel cell system. The diagram is based on values reported in the literature for the electrical conversion efficiency and cogeneration efficiency of a commercially available PAFC. Values for the fuel processor and power conditioning system are estimates. In this example, the fuel processor is approximately 85% efficient, the fuel cell stack is operating at an efficiency of 49%, and the inverter has an efficiency of 95%. The overall electrical conversion efficiency is 40%. Approximately 40% of the input energy is available as thermal energy at temperatures ranging from 40°C to 80°C (100°F to 175°F). Finally, 20% of the input energy cannot be economically recovered and is discharged to the surroundings through the exhaust gas and the power conditioner heat loss. The cogeneration efficiency reflects not only the electrical power but also the useful thermal energy available from the fuel cell system. The overall cogeneration efficiency of the example plant is 80%.

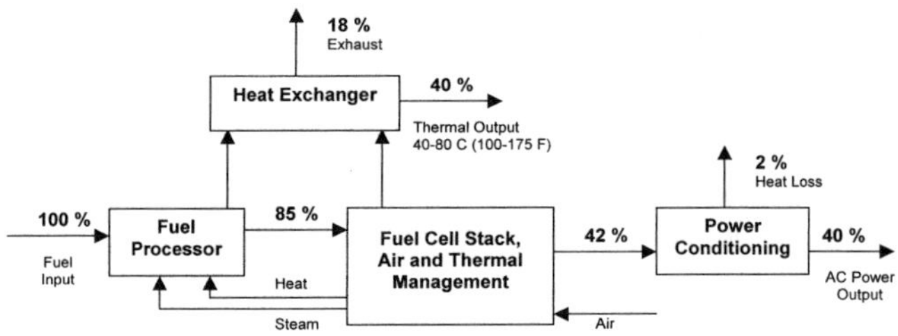

Figure 3-8 Block diagram for PAFC power plant.

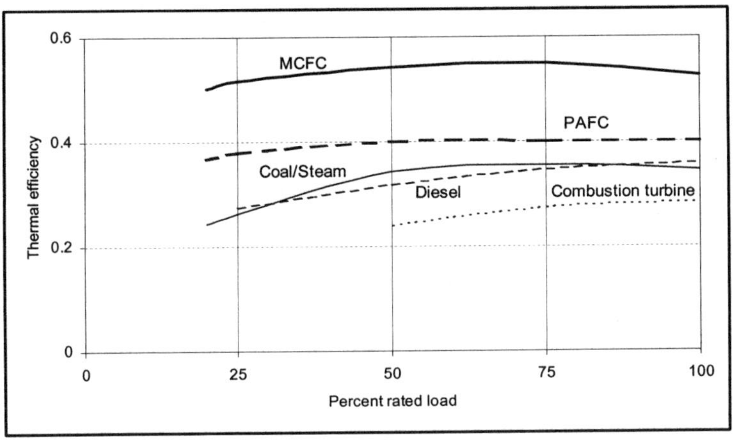

Figure 3-9 Part-load efficiency of fuel cell and conventional power systems (based on data from [4]).

Part-Load Characteristics

Fuel cell stacks generally exhibit increasing efficiency at part load as the various polarization losses, particularly ohmic losses, become smaller. On the other hand, the power consumed by the other system components, such as fans and pumps, as well as heat losses from the fuel processor, do not generally decrease directly with load. Thus, the system losses become more significant relative to the power output at part load. The net effect for the fuel cell system is a relatively flat curve relating performance to load. This is in contrast to most conventional generating systems, which operate most efficiently near full load and exhibit declining efficiency as the load decreases. Figure 3-9 compares part-load efficiencies for MCFC and PAFC systems to conventional engines. As the figure indicates, the PAFC system efficiency remains relatively constant from 40% to 100% load.

Response Time

Fuel cell response time must be characterized in two ways—cold start and hot response to load change. One of the appealing characteristics of PEMFCs in transportation applications is the fact that they can start cold in seconds. Other fuel cell technologies depend on high temperatures to maintain the operating state of the electrolyte or to facilitate the reaction kinetics. Thus, they must be heated to higher operating temperatures before they can be started. Furthermore, the rate of heating must be limited to avoid thermal stresses within the stack components. SOFCs, with their high operating temperatures and ceramic components, require particularly long cold start periods.

In stationary applications, the cold start time is less significant than in transportation applications. Since fuel cells are characterized by high first cost and low operating cost, the most economical operating strategies for fuel cell systems will likely involve operating the cells continuously. In applications where the cell remains hot continuously but experiences load changes, a more important criteria than cold start time is hot response to load change. At operating temperature, all fuel cell types respond quickly to load changes. The major factor limiting response time is likely to be the response time of the auxiliary systems, particularly the fuel processor or air compressor. Typical response times reported in the literature range from 0.3% to 10% load change per second [5]. Transients more rapid than this, such as motor starts, must be accommodated by batteries or capacitors in the power conditioning module or by connection to the power grid.

Emissions

Fuel cell stacks use fuel streams consisting of H_2 and, in some cases, CO (MCFCs and SOFCs operating on reformate gas) that have been cleaned of particulates and sulfur. Consequently, the only products emitted by the stack are H_2O and CO_2. Thus, in a fuel cell system, it is the fuel processor, not the fuel cell stack, that is the primary source of emissions. Emissions from the fuel processor depend on the type of source fuel and the type of fuel processor, but they are generally much lower than emissions from conventional combustion systems and well within air quality regulations.

Fuel cells operating on H_2 are often described as emitting only water and, thus, qualifying as zero emissions technology. This view ignores the emissions of CO_2 in the production of H_2 from a hydrocarbon source fuel. Carbon dioxide is a greenhouse gas and is considered to be a contributor to global warming. Any energy conversion technology, including fuel cells, that uses hydrocarbons as the source fuel will produce emissions of CO_2. Fuel cells, by virtue of their higher efficiency, typically use less source fuel and consequently lead to lower production of CO_2 than conventional technology.

Siting

Siting issues refer to a broad range of considerations, including size, noise, thermal emissions, and esthetics. Fuel cell stacks are modular and, in theory, can be assembled to any desired size by adding cells with minimal effect on cost and efficiency. The balance-of-plant systems are not as modular and consideration of these systems introduces economies and efficiencies of scale. Still, relative to conventional technologies, fuel cell systems are more modular and can be built in a wide range of sizes ranging from several kWe to several MWe. This allows fuel cell systems to be built in small sizes and located near the point of electricity use, thus facilitating cogeneration systems. Fuel cell systems are also relatively unobtrusive. Noise levels, which are due primarily to fans and air compressors, are comparable to those associated with residential or light commercial air-conditioning systems. The most widely commercialized PAFC design achieves a sound power level of 60 dB at 10 m [5]. Heat rejected from the fuel cell system is usually rejected directly to the air making cooling towers with visible plumes unnecessary. Commercially available systems are designed to operate unattended and are packaged for location in typical equipment spaces such as parking lots, rooftops, and basement mechanical rooms. These systems resemble other commercial building equipment such as packaged air-conditioning units. Figure 3-10 illustrates a commercially available 200 kWe PAFC fuel cell system. Figure 3-11 illustrates a prototype 100 kWe SOFC fuel cell system.

Life Expectancy

The balance-of-plant systems associated with fuel cells are expected to have a life of 20 years or more, comparable to other systems that are composed of basic equipment such as fans, pumps, heat exchangers, etc. The life of the fuel cell stack will likely be much less. The performance of most fuel cell stacks decreases with time as the catalysts are degraded and the electrodes and electrolytes are contaminated or corroded. Ultimately, the performance drops to the point that the cell

Figure 3-10 International Fuel Cells' PC25 fuel cell power plant.

(Photo courtesy of Siemens-Westinghouse Power Corporation ©2001. All rights reserved.)

Figure 3-11 Prototype 100 kWe SOFC system.

stack must be replaced. Most manufacturers of fuel cells for stationary applications have targeted 40,000 hours (roughly five years) as the minimum stack life. This goal appears attainable for all of the fuel cell technologies except for the MCFC where stack life is a primary focus of research and development efforts. At the end of its useful life, the fuel cell stack will be removed from the system and returned to the manufacturer for recycling and a new or reconditioned stack will be installed in its place. Other components that may require periodic replacement include the gas clean-up equipment and the fuel reforming catalysts.

Disposal

Economics will dictate recycling precious metal catalysts, stainless steel components, and some of the ceramic materials that are used in fuel cells. In addition, some of the compounds in the stack, such as the phosphoric acid and carbonate electrolytes, will require careful disposal procedures. Otherwise, the fuel cell system consists of piping, valves, electronic controls, and other common devices that do not present any unusual hazards during disposal. As noted by Appleby, "No special hazards or difficulties were encountered in dismantling and clearing of the New York and Tokyo Electric 4.5 MWe PAFC demonstrators [4]."

Maintainability and Availability

The fuel cell stack has no moving parts and operates at a relatively constant temperature and pressure. Therefore, the stack would be expected to have minimal maintenance requirements. Other than replacement at five-year intervals, there should be little or no on-site maintenance for the stack. The balance-of-plant contains components such as fans, pumps, piping, and controls for which maintenance intervals and procedures are well established. Figure 3-12 depicts maintenance records from an early prototype PAFC plant installed in Tokyo, Japan, in 1989 [98]. As the figure indicates, the balance-of-plant, particularly the control system, was responsible for the vast majority of maintenance problems.

As fuel cell systems mature, improved design, component selection, testing, and quality control should greatly reduce maintenance

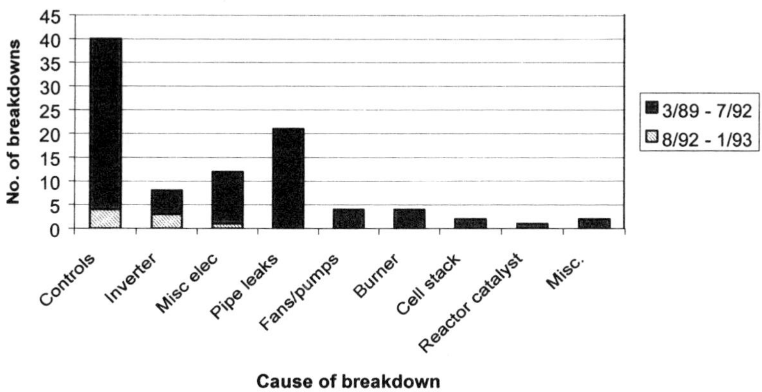

Figure 3-12 Repair history for prototype PAFC plant [98].

TABLE 3-1:
Results for (Twenty) 200 kWe Units in Field Test Ending 1995 [53]

Accumulated operating hours	
No. of units > 10,000 hours	16
Longest	> 20,000 hours
Availability	
No. of units with > 80%	13
Average availability	85%
Continuous operating hours	
No. of units > 3,000	15
Longest	6,325 (9 months)

issues associated with the balance of plant systems. In many respects, fuel cells are similar to other factory-built commercial and industrial equipment such as packaged air-conditioning equipment, absorption chillers, gas-fired boilers, etc., that are constructed from basic components. As these technologies have matured, maintainability and reliability have improved greatly. Table 3-1 presents results as of 1995 for twenty 200 kWe PAFC plants installed in Japan [53]. A similar demonstration project at U.S. Department of Defense facilities involved thirty 200 kWe PAFC plants. As of December 31, 1999, the DoD fuel cell fleet had accumulated approximately 531,000 operating hours with an unadjusted availability[1] of 68% [99]. Additional information about the DoD project is presented in Chapter 4.

Estimates of the cost for maintaining a fuel cell system range from 0.005 to 0.03 $/kWh. Roughly 40% of the respondents to the technology survey indicated that maintenance costs for MCFC and SOFC systems, including a sinking fund for stack replacement, would be less than $0.01/kWh. This is competitive with the cost of other small-scale distributed generation systems, such as natural gas engines and micro-turbines, but somewhat higher than costs associated with large steam and gas turbine systems.

1. Availability values are not adjusted for times when the fuel cell was down for periods unrelated to typical fuel cell operation (delays in maintenance personnel response, site operating conditions, etc.). Adjusting for these times would result in higher availability values [99].

Cost

Presently, only one fuel cell system is commercially available. The 200 kWe PAFC system manufactured by International Fuel Cells is currently available for roughly $5,000/kWe. Other fuel cell cost data are based on projections or targets. In general, manufacturers maintain that competitive costs are attainable provided that the market size grows large enough to support mass manufacturing technology. The most ambitious goals are associated with PEMFC technology, which is intended to compete with the internal combustion automobile engine. The goals set forth by the DOE for transportation applications specify a stack cost of $35/kWe and a system cost including fuel processor of $50/kWe. In stationary applications, fuel cell systems will be economically attractive at costs that are an order of magnitude higher than the transportation goals. A study by ADL suggests system cost targets of $2,000/kWe in order to find acceptance in applications with average utility rates and high system utilization and $1,000/kWe in order to find acceptance in broader markets [100]. Manufacturers of all fuel cell systems are making progress toward these goals through improved materials, manufacturing processes, and system design. Many foresee meeting cost objectives of $500 to $1,500 per kWe within a three to five year period.

SUMMARY OF FUEL CELL CHARACTERISTICS

Table 3-2 summarizes the fuel cell system characteristics for each of the four fuel cell technologies and compares fuel cell system characteristics to those of conventional power generating technologies.

TABLE 3-2:
Fuel Cell System Characteristics and Comparison with Conventional Generating Systems

Characteristic	PEMFC	PAFC	MCFC	SOFC	IC Engine-Generator	Micro-Turbine Generator	Gas Turbine	Steam Turbine Generator
Applicable size range, kWe	1 - 250	100-1,000	100-2,000	5-2,000	25 - 5,000	25 - 100	500 - 25,000	25 - 25,000
Projected commercialization	2001	Available	2005	2005	Available	Available	Available	Available
Projected cost (2005), $/kWe	1,000-2,000	3,000	2,000-3,000	2,000-3000	300-1,300	700 - 1,300	700 - 900	800 - 1,000
Footprint, ft^2/kWe	0.6 - 4.0				0.22 - 0.31	0.15 - 1.5	0.02 - 0.61	< 0.1
Electrical efficiency at full load, percent	40	45	45 - 50	45 - 50 PSOFC/GT: 70	25 - 45	25 - 30	25 - 40 Comb. Cycle, 40 - 60	30 - 42
Elect. eff. at ½-load, percent	40	45	45 - 50	45	23 - 40	20 - 25	20 - 35	28 - 40
Heat output, percent of input	40	35	30 - 35	35	35 - 45	50 - 55	40 - 55	Low-grade
Usable temperature, °C (°F)	50-90 (120-190)	140-200 (280-390)	600-650 (1,100-1,200)	800-1,000 (1,500-1,800)	80-480 (180-900)	200-340 (400-650)	260-600 (500-1,100)	Low-grade
Start-up time	Minutes	Hours	Hours	Hours	Seconds	Minutes	Minutes	Hours
Maintenance costs, $/kWh	0.005-0.015				0.007-0.015	0.002-0.01	0.002-0.008	0.004
Availability, percent	90 - 95				92 - 97	90 - 98	90 - 98	> 95
Expected equipment life, years	5 stack 20 bop	5 stack 20 bop	3 stack 20 bop	5 stack 20 bop	4 - 20	20	30	30

TABLE 3-2: (Continued)
Fuel Cell System Characteristics and Comparison with Conventional Generating Systems

Characteristic	PEMFC	PAFC	MCFC	SOFC	IC Engine-Generator	Micro-Turbine Generator	Gas Turbine	Steam Turbine Generator
Stack pressure, atm	1 - 3	1 - 8	1 - 10	1 - 8				
Fuel cell anode gas spec	H_2 only <10 ppm CO	H_2 only <0.5% CO <50 ppm H_2S	H_2, CO <0.5ppm H_2S	H_2, CO <1 ppm H_2S				
Fuel processing	External	External	Int. or ext.	Int. or ext.				
Possible fuels	Natural gas LPG Methanol Ethanol Gasoline Diesel	Same as PEMFC and Landfill gas	Same as PEMFC and Gasified coal	Same as PEMFC and Gasified coal	Same as PEMFC	Same as PEMFC	Same as PEMFC and Gasified coal	Same as PEMFC and Gasified coal and Solid fuels
NO_x emissions (lb/MWh)	< 0.02				2.2-2.8	0.4-2.2	0.4-4.0	1.8
References: [101], [4], [22]								

Chapter 4

FUEL CELL COGENERATION SYSTEMS IN BUILDINGS

FUNDAMENTALS OF COGENERATION SYSTEMS IN BUILDINGS

Cogeneration systems simultaneously provide both electrical and thermal energy. Cogeneration systems based on conventional generating technologies, such as reciprocating engines, gas turbines, and steam turbines, have been employed in industrial facilities, district heating systems, and, to a limited extent, in large commercial buildings. From an energy use and environmental impact point of view, cogeneration systems are very attractive. In a traditional electric power plant, only about 30% to 40% of the fuel energy is converted to electric power and the remainder is discharged to the atmosphere or a nearby river as low-grade heat. With cogeneration, this thermal energy can be recovered and put to use. Typically, an additional 30%-40% of the fuel energy can be put to use. This means that 30%-40% less energy is used and emissions of CO_2 and other pollutants can be reduced.

The widespread application of cogeneration systems in buildings has been limited by several factors.

- Conventional generating technology tends to be more economical in large sizes.
- Conventional generating technology is most efficient in large sizes and when operating near full load.
- Conventional generating technology tends to require a larger, more skilled maintenance staff than is available in many building applications.
- Noise and environmental emissions restrict the potential sites for conventional generating technology.

As we have seen, fuel cell systems overcome many of these limitations and greatly expand the possible applications for cogeneration. The following sections discuss the strategies for implementing cogeneration, the relationship between the cogeneration system and the utility grid, the economic aspects of cogeneration, and the regulatory environment affecting fuel cell cogeneration applications. Finally, several potential applications for building cogeneration are summarized.

Cogeneration Strategies

Selection of a cogeneration strategy affects the size of the cogeneration plant, the requirements for reliability, the utility interface, and the economic merits of the application. There are four basic strategies that are likely to be implemented with a fuel cell cogeneration system.

An *electrical load tracking strategy* calls for the fuel cell to have an electrical capacity that exceeds the minimum electrical requirement for the facility. The fuel cell output power increases or decreases in response to the needs of the facility. If the facility power requirements exceed the capacity of the fuel cell, then utility power must be purchased to supplement the fuel cell power. If the facility is isolated from the grid, then the fuel cell system must be capable of providing all of the necessary power for the facility, including sufficient capacity to handle transient loads such as motor starts. In addition, in many applications, the isolated system must be designed to provide redundancy so that failure of an individual fuel cell system does not result in a loss of power. With electrical load tracking strategies, the thermal output of the fuel cell system is used whenever possible, but if the heat from the cell cannot be used then it is simply rejected to the atmosphere. A separate heat source is required to provide thermal energy when the waste heat from the fuel cell is not adequate.

With a *thermal load tracking strategy*, the system is designed to follow the thermal load of the facility. Power that is generated during the course of supplying the thermal load is used by the facility to replace purchased electric power. If the electrical power exceeds the facility's needs, then the excess power can often be sold to the utility.

In contrast to tracking strategies, base load strategies call for electrical or thermal energy to be provided at a constant rate. The baseload strategy makes the most use of the fuel cell system investment. With conventional generating equipment, baseloading also allows the system to operate at peak efficiency. Since fuel cells provide excellent part-load performance, the efficiency benefit for operating at full power is not usually significant. With an *electrical base load strategy*, the fuel cell system is designed to supply the minimum amount of power required by the facility. Therefore, the fuel cell system can operate at peak power output continuously. Additional power for meeting the peak power needs of the facility is purchased from the utility. With an electrical base load strategy, a shortage of thermal energy is compensated by a separate thermal source such as a boiler. An excess of thermal energy is rejected to the atmosphere. With a *thermal base load strategy*, the system is designed to meet the minimum thermal load. Thermal energy to meet peak loads is provided by an auxiliary heat source, such as a boiler. Electrical power is purchased from or sold to the grid to balance the electrical demand for the facility with the power supplied by the fuel cell system.

Other conventional cogeneration strategies, including peak shaving and economic dispatch, are less likely to be applicable to fuel cell power systems. In *peak shaving*, the fuel cell system provides power only during on-site peaks to reduce utility demand charges or during utility peaks to avoid periods of high electric rates. Since fuel cells have a relatively high first cost and operate at high efficiencies, it is likely that once an investment is made in the fuel cell system, the most economical strategy will be to operate the system as much as possible. Thus, a peak shaving strategy that calls for fuel cell operation only during limited periods is not likely to be attractive. *Electric utility dispatched* cogeneration systems operate to provide power to a utility when the cogeneration system can operate more economically than the utility's other generating options. *Site dispatched* cogeneration systems operate only when they can provide power at a cost that is less than the marginal cost at which electricity can be purchased from the utility. Economically dispatched fuel cell systems, like peak shaving systems, may not provide sufficient operating hours to justify the purchase price of the system.

Economic Considerations

The decision to install a fuel cell cogeneration system will almost always be based on economic considerations. In some cases, the economic impact associated with a fuel cell system will be attributable to improvements in power quality. In most cases, however, the savings will result from reduced utility costs.

There are a variety of techniques for evaluating the economic merits of a fuel cell cogeneration system, ranging from simple to complex. The simple techniques can be implemented with minimal data and calculations but give results that can be misleading and that must be interpreted with caution. More complex techniques require extensive data collection and numerous calculations but can yield more accurate results. Furthermore, the more complex techniques require the analyst to think through the application in greater detail, forcing a more thorough understanding of the assumptions and interrelationships inherent in the analysis. Here, two representative types of analyses are discussed. First, a simple cost of electricity calculation is described. Then a more complete hour-by-hour approach for estimating annual utility cost is discussed. The simple approach is then applied to give representative results for residential and commercial applications.

Simple Cost of Electricity Analysis.

The simplest type of analysis consists of comparing the cost of electricity generated by the cogeneration system to the cost of electricity available from the utility. The cost of cogenerated electricity is estimated by determining capital, maintenance, and utility costs on a $/kWh basis. The electricity cost is reduced by the value of the thermal energy produced. These calculations are reflected in Equation 4-1.

$$COE = \frac{(A/P,\,n,\,r) \times CC}{EFLH \times FP} + \frac{C_1 \times FC}{\eta_E} + MC - \frac{C_1 \times F_1 \times \varepsilon_T \times FC}{\eta_E \times \eta_A} \qquad (4\text{-}1)$$

where

$(A/P,n,r)$	=	annual cost factor for a fuel cell life of *n* years and an interest rate, *r*
CC	=	capital cost of the fuel cell system ($)
$ELFH$	=	number of equivalent full-load hours, which corresponds to the number of hours at full power that would be required to produce the amount of energy supplied by the system during the year (hours)
FP	=	rated or full-load power of the fuel cell system (kW)
C_1	=	conversion factor relating the units for the fuel energy to the units for electrical energy (e.g., 3.412 MBtu/kWh)
FC	=	fuel cost ($/MBtu)
MC	=	maintenance cost ($/kWh)
η_E	=	average electrical conversion efficiency of the fuel cell system
η_A	=	efficiency of the thermal source displaced by the thermal energy of the fuel cell system
F_1	=	thermal use factor that reflects the fraction of thermal energy from the fuel cell system that can be used within the facility
ε_T	=	fraction of the fuel energy supplied to the fuel cell that is available as usable thermal energy from the system

In this expression, the first term represents the annualized cost of the fuel cell spread over the annual amount of electricity produced. The number of periods, n, should correspond to the life of the balance of plant system. The stack life will be shorter, but the cost of stack replacement will be treated as a maintenance cost. The second term is the fuel cost per unit of electricity produced. The third term is the maintenance cost per unit of electricity produced. This term must include routine annual maintenance costs plus a sinking fund to periodically replace the fuel cell stack. The final term reflects the value of the thermal energy that would have been required from an alternative system, such as a boiler, but that is instead provided as a by-product of fuel cell operation. The amount of thermal energy provided by the fuel cell may exceed that which is required by the facility. Or, the need for thermal energy may not coincide with the availability of thermal energy from the fuel cell. The factor, F_1, indicates the average fraction of thermal energy from the fuel cell system that can actually be used to replace energy from other sources. The value of F_1 cannot exceed 1 and is further restricted by

$$F_1 \leq \frac{r_{TE}\eta_E}{\varepsilon_T} \qquad (4\text{-}2)$$

where r_{TE} is the annual average thermal electric ratio for the building.

The cost of electricity, as calculated by Equation 4-1, can be compared to the average cost of electricity for the facility to determine whether electricity can be provided more economically with the fuel cell system.

This simplified approach to evaluating a fuel cell cogeneration system has two major drawbacks. First, the factor, F_1, which determines the extent to which thermal energy is recovered, is difficult to estimate. A careful determination of this factor would require hour-by-hour knowledge of the electrical and thermal energy profiles. Furthermore, incorporation of this information would greatly complicate the approach.

Another problem with the simplified approach is the determination of the cost of electricity supplied by the utility. An average cost of electricity can be calculated based on annual cost and annual energy use in kWh. However, this cost is really only applicable when comparing the option of supplying all of the electricity on-site to purchasing all of the electricity from the utility. In other cases, when only a fraction of the electricity is generated on-site, the challenge is to properly evaluate the cost of utility-furnished electricity that is replaced by on-site electricity. Depending on the utility pricing structure, this cost can vary with the time of day, the season of the year, and the electricity demand and energy use history for the facility. The cost of electricity can certainly be determined, but doing so may require a very detailed calculation that incorporates hour-by-hour electricity data. Thus, the "simplified approach" actually only looks simple because the complexity is hidden in the values of the thermal use factor F_1 and the average electricity cost. The method is only as accurate as the determination of these values.

Hourly System Modeling.

In hourly system modeling, the electrical and thermal loads are determined for each hour of the day for each unique day of the year. At each hour, the amount of electricity generated by the fuel cell system and the amount of electricity purchased from the utility are determined based on the system operating strategy. In addition, the amount of heat from the fuel cell that is needed to satisfy the thermal load is determined. If the fuel cell efficiency varies significantly with load, this effect can also be reflected in the calculations. Finally, the prices of gas and electricity are determined based on the actual utility pricing structures for each hour if necessary. The hourly approach not only provides more accurate results but also permits testing different operating strategies in order to minimize the cost of energy.

The electric utility can usually provide records of hourly power requirements for most large commercial and industrial facilities. Hourly estimates of the thermal load are more difficult to obtain. It may be necessary to model potable hot water use based on empirical data and to estimate space heating requirements using energy estimating software. Results from these estimates can then be reconciled with the actual monthly bills to yield more accurate results. In other cases, particularly in industrial plants with significant process heating loads, it may be necessary to install temporary metering to determine the thermal energy use.

Once hourly data are known, this information is combined to determine the annual cost of operating the fuel cell cogeneration system and the annual cost of purchasing the energy that would be produced by the fuel cell cogeneration system. The difference between these values reflects the net annual savings associated with the fuel cell system. These savings can be compared to the first cost of the system on the basis of an appropriate economic criterion. Common criteria include cost to benefit ratio, internal rate of return, and life-cycle cost.

Calculation of the annual savings, *AS*, associated with the fuel cell system can be accomplished by application of the following equation:

$$AS = R[EL_{i=1\ldots I}] - R[(EL_i - EP_i)_{i=1\ldots I}]$$
$$+ \sum_{i=1}^{I} \frac{C_1 \times FC_i}{\eta_A} \times min\langle \frac{\varepsilon_T \times EP_i}{\eta_E}, TL_i \rangle \qquad (4\text{-}3)$$
$$-MC \sum_{i=1}^{I} EP_i - \sum_{i=1}^{I} \frac{C_1 \times FC_i \times EP_i}{\eta_E}$$

where

$R[...]$ = a function described by the electricity rate structure. Given a series of electrical energy requirements $EL_{i=1\ldots I}$ for each hour of the year, the rate structure function can be used to calculate the resulting annual electricity cost.

EL_i = electrical energy requirement (load) for each hour, *i*, of the year.

EP_i = electrical energy output from the fuel cell system for each hour, *i*, of the year.

FC_i = the fuel cost for each hour, *i*, of the year

TL_i = thermal requirement (load) for each hour, *i*, of the year.

Other variables are as defined following Equation 4-1.

The first term of Equation 4-3 is the annual cost of electricity based on the hourly power requirements EL_i, when the fuel cell system is not purchased. The second term is the annual cost of electricity based on the hourly power requirement less the power supplied by the fuel cell system. The difference between the first and second terms represents the annual electricity cost savings. The third term reflects the fuel cost savings associated with using thermal energy from the fuel cell system. Within this term, the first expression in braces is the thermal energy available from the fuel cell system based on the electricity produced. The second expression in braces is the thermal energy required by the application. The minimum of these values represents the actual thermal energy from the fuel cell that is used in the application. Multiplying this value by the cost of producing this thermal energy with the alternate thermal system (e.g., a boiler) gives the savings associated with the fuel cell cogeneration system for each hour of the year. Summing over the year yields the annual fuel cost savings associated with the cogeneration system. The third term in Equation 4-3 represents the annual maintenance cost associated with the fuel cell system. This term is equal to the product of the total amount of electricity provided by the fuel cell system and the maintenance cost per kWh. The final term in Equation 4-3 represents the annual cost of fuel for the fuel cell system.

While modern spreadsheet programs can accomplish the calculations required by Equation 4-3 for each of the 8,760 hours of the year, it is often easier to recognize that many days of the year have similar electrical profiles. For example, during a particular month, the electrical profile for each weekday may be the same. A different profile may characterize each weekend day. Thus, the month can be characterized by two days with the results from each typical day weighted by the number of occurrences of the typical day. Thus, the entire year could be described by 12 months, each containing 2 days, each consisting of 24 hours, for a total of 576 hours. This approach may significantly reduce the amount of data that must be handled without significantly affecting the results.

Table 4-1 illustrates an excerpt of a spreadsheet used to perform the calculations indicated by Equation 4-3. In this example, the electricity rate structure was relatively simple, with an energy cost of $0.03/kWh and a monthly demand charge of $12/kWe. The energy savings associated with the fuel cell for the month of May are found to be $24,419.

RESIDENTIAL COGENERATION APPLICATIONS

Residential energy use is characterized by a low electrical load factor and a relatively high thermal to electric ratio. The thermal use factor varies with the season. During the heating season, virtually all of the thermal energy from the fuel cell system can be used for heating. During the summer, electricity use is high and heating needs are low so that only a small fraction of the thermal energy produced by the system can be used. Both thermal and electrical requirements tend to be intermittent and it is likely that some type of thermal storage will be required to make residential systems economical.

Figure 4-1 illustrates the economics associated with a residential cogeneration system operating with an isolated (grid independent) electrical load following strategy. In this figure, the cost of natural gas is indicated on the horizontal axis. The resulting cost of electricity is indicated on the vertical axis for various initial system costs. The cost of electricity was determined using the simple cost of electricity analysis reflected in Equation 4-1 and the assumptions indicated in the figure.

For comparison purposes, representative residential electricity and gas rates are depicted for various regions of the country [102]. For example, in the South, electricity and gas are relatively cheap at

TABLE 4-1a:
Economic Analysis of Fuel Cell Cogeneration System—Hourly Data

Date	Time	Elect. Power	Therm Load	Energy Sources		
				FC Power	Utility Power	Useful Heat
		kW	kW	kW	kW	kW
1-May	8	2638	836	1000	1638	778
	9	2644	880	1000	1644	778
	10	2707	748	1000	1707	748*
	11	2580	792	1000	1580	778
	12	2586	704	1000	1586	704*
	13	2517	748	1000	1517	748*
	14	2678	792	1000	1678	778
	15	2655	660	1000	1655	660*
	16	2638	748	1000	1638	748*
	17	2661	704	1000	1661	704*
	18	2650	792	1000	1650	778

* Indicates useful thermal energy is limited to that required by the thermal load.

TABLE 4-1a: (Continued)
Economic Analysis of Fuel Cell Cogeneration System—Hourly Data

				Energy Sources		
		Elect Power	Therm Load	FC Power	Utility Power	Useful Heat
Date	Time	kW	kW	kW	kW	kW
	19	2742	836	1000	1742	778
	20	2782	660	1000	1782	660*
	21	2696	792	1000	1696	778
	22	2748	704	1000	1748	704*
	23	2609	704	1000	1609	704*
	24	2494	792	1000	1494	778
2-May	1	2661	528	1000	1661	528*
	2	2650	880	1000	1650	778
	3	2673	616	1000	1673	616*
	4	2678	704	1000	1678	704*
	5	2661	704	1000	1661	704*
	6	2661	792	1000	1661	778
	7	2592	748	1000	1592	748*
Similar Results for Other Hours in May						
Totals for May		1971600		744000	1227600	523776

* Indicates useful thermal energy is limited to that required by the thermal load.

Table 4-1b:
Economic Analysis of Fuel Cell Cogeneration System—Calculation of Savings for May

	Monthly Results *without* Fuel Cell Cogeneration	**Monthly Results *with* Fuel Cell Cogeneration**
Peak demand, kW	2,782	1,782
Electrical energy purchased, kWh	1,971,600	1,227,600
Electrical demand cost	$33,384	$21,384
Electrical energy cost	$88,722	$55,242
Total electricity cost	$122,106	$76,626
Net electricity savings		$45,480
Fuel cell natural gas use, MCF		5,638
Boiler fuel savings from heat recovery, MCF		(2,233)
Net natural gas use, MCF		3,405
Net natural gas expense		($13,621)
Net utility savings (electricity and gas)		$31,859
Maintenance expense		($7,440)
Net savings		**$24,419**

Basis for calculations:

Fuel cell electrical conversion efficiency	0.45
Fuel cell thermal power fraction	0.35
Boiler efficiency	0.80
Electrical power demand cost	$12/kW
Electrical energy cost	$0.045/kWh
Natural gas cost	$4/MCF
Fuel cell maintenance cost	$0.01/kWh

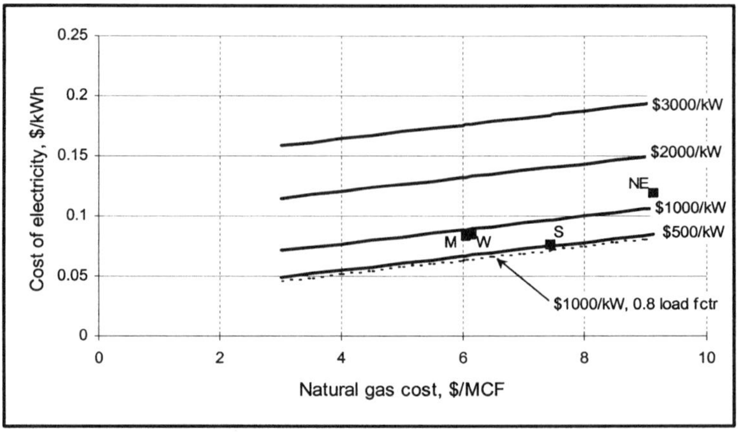

Assumptions:

Electrical load factor:	0.35 (except where noted)
Thermal/electric ratio:	2.2
Thermal use factor:	0.65
Efficiency of alternate heat source:	0.8
Electrical conversion efficiency:	0.4
Thermal output fraction:	0.4
System maintenance cost:	$0.01/kWh
Expected system life:	20 years
Rate of return:	12 percent

Figure 4-1 Cost of electricity for fuel cell cogeneration in a residence.

$0.076/kWh and $7.44/MCF, respectively. In contrast, in the Northeast, the average electricity cost is $0.119/kWh and the average gas cost is $9.13/MCF.

Several conclusions can be drawn from Figure 4-1. First, fuel cell system costs will have to drop to roughly $1,000/kWe before they can begin to compete with utility power in most regions of the country. Second, fuel cell systems are likely to initially be competitive in regions of the country such as the Northeast where electricity costs are high. Finally, in Figure 4-1, the dashed line indicates the cost of electricity for the same assumptions as the other cases except that the load factor has been increased to 0.8. As indicated, this significantly decreases the cost of electricity. This case would correspond to a grid-connected electrical load following strategy where power from the grid is used to help meet peak electrical requirements. This case might also correspond to a residence that uses thermal storage or a cooling system other than electric vapor compression air conditioning.

COMMERCIAL COGENERATION APPLICATIONS

Energy use in commercial buildings varies with the type of activity in the building. Electrical load factors range from a low of 0.09 for places of worship to 0.5 for food sales applications. Thermal to electric ratios range from 0.2 for food sales applications, where electric refrigerating equipment uses considerable energy, to 2.59 for places of worship, which use little electric power when unoccupied but must be heated throughout the winter. The most promising applications are likely to be lodging and health care applications where load factors are in the range of 0.25 to 0.35 and thermal to electric ratios are 1 or higher. These applications also typically require backup power systems for operating lights and emergency equipment. A fuel cell cogeneration system that provides electricity, thermal energy, and backup power is likely to be one of the first systems to be economical in building applications. Other early adopters of fuel cell system technology in the commercial sector are likely to include data processing, telecommunications, and high tech manufacturing applications, such as integrated circuit fabrication plants, where the need for power with high quality and reliability often necessitates an auxiliary power system. A summary of commercial applications is provided at the end of this chapter.

Figure 4-2 illustrates the economics associated with a commercial cogeneration system operating with an isolated (grid independent) electrical load following strategy. In this figure, the cost of natural gas is indicated on the horizontal axis. The resulting cost of electricity is indicated on the vertical axis for various initial system costs. The cost of electricity was determined using the simple cost of electricity analysis reflected in Equation 4-1. The analysis is based on the indicated assumptions that are characteristic of a large office building with a floor area of roughly 46,500 m^2 (500,000 ft^2). The 12% rate of return corresponds roughly to a 7.5-year payback period. A payback period of 7.5 years is likely to be the maximum that would be considered in a commercial application.

For comparison purposes, representative electricity and gas rates are depicted for various regions of the country [103]. For example, in the South, electricity and gas are relatively cheap at \$0.06/kWh and \$4.85/MCF, respectively. In contrast, in the Northeast, energy costs are higher with electricity averaging \$0.1/kWh and gas \$5.85/MCF.

Several conclusions can be drawn from Figure 4-2. As in the case of residential systems, fuel cell system costs will have to drop to nearly \$1,000/kWe before they can begin to compete with utility power in most regions of the country. Second, fuel cell systems are likely to initially be

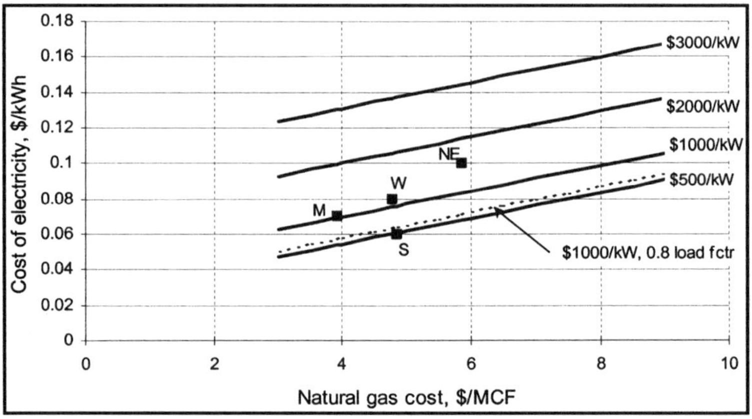

Assumptions:

Electrical load factor:	0.5 (except where noted)
Thermal/electric ratio :	0.5
Thermal use factor:	0.3
Efficiency of alternate heat source:	0.8
Electrical conversion efficiency:	0.4
Thermal output fraction:	0.4
System maintenance cost:	$0.01/kWh
Expected system life:	20 years
Rate of return:	12 percent

Figure 4-2 Cost of electricity for fuel cell cogeneration in an office building.

competitive in regions of the country such as the Northeast where electricity costs are relatively high. Finally, in Figure 4-2, the dashed line indicates the cost of electricity for a fuel cell system cost of $1,000/kWh and the same assumptions listed above except that the load factor has been increased to 0.8. A load factor of 0.8 or higher would be characteristic of a grid-connected electrical load following strategy where the fuel cell is not sized to meet peak power but rather an average power requirement. As indicated, this decreases the cost of electricity significantly. However, in this case, the cost of the electricity provided by the utility must be considered carefully. The fuel cell is displacing electricity that is purchased as part of the facility's base electrical load. The cost of electricity must therefore be compared to the rate for baseload electricity, which is likely to be less than the average electricity rate.

REGULATORY ISSUES

Installation of fuel cell cogeneration systems in buildings will require the approval of the appropriate building authorities. Approval is typically granted based on compliance with building codes and standards. Thus, before fuel cell cogeneration systems can be widely implemented, codes and standards must be developed addressing this new technology. There are several organizations working to develop these codes and standards and some have already been published. The primary organizations involved in this effort are:

- International Code Council (ICC)—develops model building codes that are adopted by federal, state, and local authorities. The ICC international code replaces the three separate model codes previously published by Building Officials and Code Administrators International (BOCA), International Conference of Building Officials (ICBO), and Southern Building Code Congress International (SBCCI).
- National Fire Protection Organization (NFPA)—develops standards for preventing fire and loss of life. NFPA standards are adopted by reference in the model building codes.
- American Society of Mechanical Engineers (ASME)—develops test standards and design codes such as the ASME Boiler and Pressure Vessel Code. ASME standards are adopted by reference in the model building codes.
- American National Standard Institute (ANSI)—develops and maintains national standards for equipment and procedures.
- Institute of Electrical and Electronic Engineers (IEEE)—develops standards for the design of electrical and electronic equipment.
- National Hydrogen Association—develops standards for the safe design and construction of hydrogen fuel systems.
- International Standards Organization (ISO)—develops international standards in a variety of areas including safety, quality control, and measurements.

Other participants in the codes and standards development process include organizations such as Underwriters Laboratory and National Evaluation Services, which test and evaluate equipment for compliance to product standards.

Codes and standards for fuel cell systems that are currently published or under development by these organizations are described below.

2000 International Mechanical Code (IMC). This code has been published and is available for adoption by federal, state, and local building authorities. Section 924 of the IMC mandates that fuel cell systems with a power output not exceeding 1 MWe must be tested in accordance with ANSI Z21.83 and installed in accordance with the manufacturers' requirements. Larger fuel cell systems must be documented to be equivalent from a safety and health standpoint to other technologies accepted in the code.

ANSI Z21.83-1998/CSA 12.10: Fuel Cell Power Plants. This standard, developed by IAS/CSA and adopted by ANSI, was published in 1998. The standard applies to packaged fuel cell power plants with capacities up to 1 MWe operating on natural gas or liquefied petroleum (LP) gas. The standard provides construction, performance, and safety criteria.

NFPA 853: Standard for the Installation of Stationary Fuel Cell Power Plants. This standard was published in 2000 and applies to stationary fuel cell power plants larger than 50 kWe including packaged and field-constructed systems. The standard addresses siting, fuel supplies, ventilation, and fire protection.

NFPA 70: National Electric Code (NEC). The NEC is the primary standard governing the installation of electrical systems in buildings. A new article, Article 691, covering packaged fuel cell systems has been proposed and accepted in principle. Article 691 would address installation, wiring, circuit protection, and connection to other systems. Action on Article 691 is expected by May 2001.

ASME PTC-50: Performance Test Code for Fuel Cell Power System Performance. PTC-50 is currently under development and is scheduled for publication in 2002. The test code introduces definitions, and provides test methods and procedures for characterizing the performance of fuel cell power systems.

IEEE P1547: Standards for Distributed Resources Interconnection with Electric Power Systems. P1547 is currently under development and is scheduled for completion in 2001. The standard is intended to establish uniform criteria for interconnecting distributed generation systems, including fuel cells, photovoltaics, and wind generators, with the electric utility grid. The standard will establish interconnect classifications, performance and safety criteria, and procedures for operation, safety, testing, and maintenance of the interconnection.

As a new technology, fuel cell cogeneration systems are likely to be initially viewed with caution by the building safety community. Recognition of fuel cell cogeneration systems in codes and standards will facilitate the introduction of this technology.

BUILDING COGENERATION DEMONSTRATION PROJECT RESULTS

The U.S. Department of Defense (DOD) and the U.S. Army Construction Engineering Research Laboratory have conducted a demonstration program in which thirty 200 kWe PAFC fuel cell systems fueled by natural gas have been installed in a variety of building applications including the following:

- Boiler plants (11 installations)
- Hospitals (7 installations)
- Gymnasiums/pools (3 installations)
- Dorms/barracks (3 installations)
- Office buildings (3 installations)
- Laundry (1 installation)
- Kitchen (1 installation)
- Control center (1 installation)

Information obtained from this demonstration program includes (for each installation) thermal utilization, energy savings, emissions reductions, equipment availability, and operating experience. Thermal utilization was lowest (9% to 19%) in office buildings and a galley and highest (100%) in boiler plants. Likewise, estimated annual energy savings from the 200 kWe fuel cell systems ranged from $16,000 in an office building application to $103,000 in a boiler plant application. As of December 31, 1999, the total fleet of systems had operated for 531,000 hours and generated 91,720 MWh of electricity with an unadjusted availability rate of 68%. More detailed information on each application is available from the program website [99].

BUILDING APPLICATION REQUIREMENTS

The feasibility of building cogeneration depends on the characteristics of the particular building application. Cogeneration is most likely to be attractive when the application has a high load factor, a thermal/electric ratio of one or more, and a high thermal use factor. Cogeneration may also be attractive when a second power system is required for some other reason, such as emergency power for life safety or backup power for a critical data processing or manufacturing operation. In this case, the first cost of the fuel cell system is partially offset by the cost of the conventional system that would have been purchased. The following sections summarize the essential requirements that must be considered when evaluating fuel cell cogeneration systems for the following:

- Office buildings
- Lodging
- Food sales
- Health care
- Schools
- Mercantile

For each application, information from the 1995 Commercial Building Energy Consumption Survey [103] is used to estimate the typical power requirement and electrical load factor. In addition, the survey data are used to determine the average annual thermal/electric ratio. Power quality, temperature requirements for cogeneration, load strategy, fuel source, and emissions are addressed qualitatively. Typical codes and standards that govern each application are also identified. Finally, the most likely maintenance arrangements are identified based on industry practices for each application. In general, large buildings will have in-house maintenance staff that can service balance of plant components such as pumps, fans, and compressors. Maintenance of components that are specific to the fuel cell or fuel processor will likely be covered under service contracts with the system manufacturer. Smaller facilities will rely entirely on service contracts with the manufacturer.

Office Buildings

Typical Size Range. Electrical power demand is typically in the range of 3 to 8 W/ft^2. Lights and appliances typically account for roughly half of the electrical demand. Thus, there is a substantial electrical demand whenever the building is occupied.

Electrical Load Factor. Since many office buildings are occupied primarily during a 40- to 50-hour work week, the electrical load factor is relatively small. Typical electrical load factors are in the range of 0.20 to 0.35.

Power Quality and Reliability. For most office buildings, power quality is important, but backup power systems are not common. Exceptions include data processing and telecommunication centers where loss of power, even briefly, can lead to substantial expenses or loss of revenue. In addition, in larger buildings, backup power systems are required to operate emergency lighting and life-safety equipment including elevators, fire pumps, and smoke management systems.

Thermal/Electrical Ratio. Typical thermal to electric ratios (based on annual energy use) range from approximately 12% if thermal energy is used only for water heating to as high as 50% if used for water heating and space heating.

Temperature. Thermal energy available at temperatures of 80°C to 90°C (175°F to 195°F) can be used for water heating, space heating, and single-stage absorption cooling. Higher grade thermal energy at temperatures of 120°C to 190°C (250°F to 375°F) can be used for these services as well as for two-stage absorption cooling cycles and for supplying existing steam networks. The availability of thermal energy at temperatures exceeding 190°C (375°F) is not an advantage unless steam at pressures exceeding 10 atm (130 psig) is required for large distribution systems.

Load Strategy. The most economical load strategy is likely to be a baseload electrical strategy in which the fuel cell is sized to meet the building's baseload requirements and then operated continuously. Equipment sized to meet the peak building power requirement is not likely to be economical unless power can be sold back to the utility at attractive rates or an independent power system is required for power quality reasons.

Fuel Source. The most economical fuel is natural gas. Composition of the gas and use of additives and odorants should be verified and reviewed with the fuel cell manufacturer to ensure proper operation of the fuel processor. For backup purposes, other fuels such as propane can be used but may require modifications to the fuel processor.

Codes and Standards Compliance. The fuel cell system should be installed in accordance with local codes. A preliminary design review with code administrators is advisable since the technology is relatively new. Codes and standards govern the fuel cell itself, connection to the gas utility, connection to the electric utility and the building's electrical system, and location of the fuel cell. Codes and standards that are likely to be applicable include:

International Mechanical Code – 2000
NFPA 853
ANSI Z21.83-1998
ASME PTC 50 (under development)
IEEE P1547
NEC

Emissions. In general, fuel cells can meet the most stringent emission requirements and are viewed as environmentally beneficial technologies. However, fuel cell installations will generally require a permitting procedure similar to that required for other gas-fired appliances such as boilers, furnaces, and engines.

Maintenance. In small facilities, the maintenance staff is likely to be limited. Service contracts will be required to ensure that routine maintenance is performed. In larger buildings, maintenance staff is likely to be familiar with the service needs of the components making up the balance-of-plant systems.

Lodging

Typical Size Range. Electrical power demand is typically in the range of 3 to 8 W/ft^2. Lights and appliances account for a large part of the electrical demand. Thus, there is a substantial base load electrical demand whenever the building is occupied.

Electrical Load Factor. Full service hotels with convention centers and dining are occupied for business during the day and lodging at night. Hotels offering only lodging tend to have more variable occupancy and consequently lower load factors. Typical electrical load factors for all lodging facilities range from 0.28 to 0.44.

Power Quality and Reliability. For most hotels, power quality is not an issue. In larger buildings, backup power systems may be used to operate emergency lighting and life-safety equipment including elevators, fire pumps, and smoke management systems. Fuel cells installed to serve these requirements may be operated continuously to provide power and thermal energy.

Thermal/Electrical Ratio. Typical thermal/electrical ratios (based on annual energy use) range from approximately 1.0 if thermal energy is used only for water heating to as high as 1.4 if used for water heating and space heating. The high water heating demand makes cogeneration attractive. Installation of hot water storage tanks allows heat to be recovered when it is available from the fuel cell system.

Temperature. Thermal energy available at temperatures of 80°C to 90°C (175° to 195°F) can be used for water heating, space heating, and single-stage absorption cooling. Higher grade thermal energy at temperatures of 120°C to 190°C (250°F to 375°F) can be used for these services as well as for two-stage absorption cooling cycles and for supplying existing steam networks. The availability of thermal energy at temperatures exceeding 190°C (375°F) is not likely to be an advantage.

Load Strategy. The most economical load strategy is likely to be a baseload electrical strategy in which the fuel cell is sized to meet the building's baseload electrical requirements and then operated continuously. Another possible strategy is to provide hot water storage to establish a relatively constant water heating requirement and then size the fuel cell system to meet the average water heating needs. Equipment sized to meet the peak building power requirement is not likely to be economical unless power can be sold back to the utility at attractive rates.

Fuel Source. See discussion for office buildings.

Codes and Standards Compliance. See discussion for office buildings.

Emissions. See discussion for office buildings.

Maintenance. In small facilities, the maintenance staff is likely to be limited. Service contracts will be required to ensure that routine maintenance is performed. In larger hotels, maintenance staff is likely to be familiar with the service needs of the components making up the balance-of-plant systems.

Food Sales

Typical Size Range. Electrical power demand is typically in the range of 10 to 22 W/ft^2. Refrigeration equipment and lights account for a large part of the electrical demand.

Electrical Load Factor. Since many stores are open 24 hours per day, the electrical load factor is relatively high. Typical electrical load factors range from 0.37 to 0.59.

Power Quality and Reliability. Many stores have backup power systems to keep refrigerating units in operation in the event of a power failure. Fuel cells installed to serve these requirements may be operated continuously to provide power and thermal energy.

Thermal/Electrical Ratio. Typical thermal electric ratios (based on annual energy use) range from approximately 0.05 if thermal energy is used only for water heating to as high as 0.21 if used for water heating and space heating. Thus, thermal energy goes essentially unused during the summer and use is limited even in the winter.

Temperature. Thermal energy available at temperatures of 80°C to 90°C (175°F to 195°F) can be used for water heating and space heating. Some stores use desiccant systems for humidity, and high-temperature thermal energy could conceivably be used in conjunction with these systems.

Load Strategy. The most economical load strategy is likely to be a baseload electrical strategy in which the fuel cell is sized to meet the building's baseload requirements and then operate continuously. Alternately, the fuel cell system may be sized to meet the backup power requirements of the refrigerating equipment and then operate continuously at full power with the electric utility providing supplemental power. Equipment sized to meet the peak building power requirement is not likely to be economical unless power can be sold back to the utility at attractive rates.

Fuel Source. See discussion for office buildings.

Codes and Standards Compliance. See discussion for office buildings.

Emissions. See discussion for office buildings.

Maintenance. Maintenance will likely be performed under service contracts.

Health Care

Typical Size Range. Electrical power demand is typically in the range of 4 to 10 W/ft^2. Lights, appliances, and HVAC equipment make up the majority of the electrical demand.

Electrical Load Factor. Medical office buildings and outpatient clinics have electrical load profiles that are similar to office buildings. Primary health care centers have more constant electrical loads with variations that are attributable mostly to the air-conditioning load. As a group, health care facilities have load factors ranging from 0.2 to 0.4.

Power Quality and Reliability. For medical office buildings, power quality issues are similar to those for other office buildings. For primary health care centers, power quality and reliability are critical. In these facilities, backup power systems are required to operate medical equipment, emergency lighting, and life-safety equipment, including elevators, fire pumps, and smoke management systems. Outpatient clinics and nursing homes have requirements that fall somewhere between those for office buildings and primary health care centers as determined by the specific use of the facility. Fuel cells installed to meet emergency and backup power requirements may be operated continuously to provide power and thermal energy.

Thermal/Electrical Ratio. Typical thermal/electrical ratios (based on annual energy use) range from approximately 0.50 if thermal energy is used only for water heating to as high as 1.0 if used for water heating and space heating.

Temperature. Thermal energy available at temperatures of 80°C to 90°C (175°F to 195°F) can be used for water heating, space heating, and single-stage absorption cooling. Higher grade thermal energy at temperatures of 120°C to 190°C (250°F to 375°F) can be used for these services as well as for two-stage absorption cooling cycles and for supplying existing steam networks. The availability of thermal energy at temperatures exceeding 190°C (375°F) is not likely to be an advantage except in large medical centers with central steam systems.

Load Strategy. In general, the most economical load strategy is likely to be a baseload electrical strategy in which the fuel cell is sized to meet the building's baseload requirements and then operate continuously. For buildings with large backup power requirements, the most economical strategy may be to size the fuel cell system to meet the backup power requirements. The system would then be run continuously with utility power supplying the rest of the load. Another possible strategy is to size the fuel cell based on the thermal energy requirements for water heating.

Fuel Source. See discussion for office buildings. On-site storage of propane for a backup fuel may be required by code.

Codes and Standards Compliance. See discussion for office buildings. In addition, the Life Safety Code, and specific health care codes should be consulted for requirements for backup power systems.

Emissions. See discussion for office buildings.

Maintenance. In large medical centers, the maintenance staff is likely to be familiar with the components comprising the balance-of-plant systems and be capable of performing routine maintenance.

Schools

Typical Size Range. Electrical power demand is typically in the range of 2 to 9 W/ft^2. Lights and appliances account for a large part of the electrical demand. Thus, there is a substantial electrical demand whenever the building is occupied.

Electrical Load Factor. Since many schools are occupied primarily during a 40- to 50-hour week and may be sparsely occupied during the summer, the electrical load factor is relatively small. A typical electrical load factor is 0.2.

Power Quality and Reliability. For most schools, power quality is not an issue. In larger buildings, backup power systems may be used to operate emergency lighting and life-safety equipment, including elevators, fire pumps, and smoke management systems. Fuel cells installed to serve these requirements may be operated continuously to provide power and thermal energy.

Thermal/Electrical Ratio. Typical thermal/electrical ratios (based on annual energy use) range from approximately 0.60 if thermal energy is used only for water heating to as high as 1.7 if used for water heating and space heating.

Temperature. Thermal energy available at temperatures of 80°C to 90°C (175°F to 195°F) can be used for water heating, space heating, and single-stage absorption cooling. Higher grade thermal energy at temperatures of 120°C to 190°C (250°F to 375°F) can be used for these services as well as for two-stage absorption cooling cycles and for supplying existing steam networks. The availability of thermal energy at temperatures exceeding 190°C (375°F) is not likely to be an advantage.

Load Strategy. The most economical load strategy is likely to be a baseload electrical strategy in which the fuel cell is sized to meet the building's baseload requirements and then operated continuously. Equipment sized to meet the peak building power requirement is not likely to be economical unless power can be sold back to the utility at attractive rates.

Fuel Source. See discussion for office buildings.

Codes and Standards Compliance. See discussion for office buildings.

Emissions. See discussion for office buildings.

Maintenance. Maintenance staff is likely to be limited and maintenance will likely be accomplished through service contracts. In school systems, it may be possible to share maintenance expertise among schools.

Mercantile

Typical Size Range. Electrical power demand is typically in the range of 2 to 9 W/ft^2. Lights and appliances account for a large part of the electrical demand. Thus, there is a substantial electrical demand whenever the building is occupied.

Electrical Load Factor. Since most retail facilities are occupied primarily during a 50- to 60-hour week, the electrical load factor is relatively small. Typical load factors range from 0.16 to 0.33.

Power Quality and Reliability. In most cases, power quality is not an issue. In larger enclosed shopping malls, backup power systems may be used to operate emergency lighting and life-safety equipment, including elevators, fire pumps, and smoke management systems. Fuel cells installed to serve these requirements may be operated continuously to provide power and thermal energy.

Thermal/Electrical Ratio. Typical thermal/electrical ratios (based on annual energy use) range from approximately 0.40 if thermal energy is used only for water heating to as high as 0.9 if used for water heating and space heating.

Temperature. Thermal energy available at temperatures of 80ºC to 90ºC (170ºF to 190ºF) can be used for water heating, space heating, and single-stage absorption cooling. Higher grade thermal energy at is not likely to be an advantage.

Load Strategy. The most economical load strategy is likely to be a baseload electrical strategy in which the fuel cell is sized to meet the building's electrical baseload requirements and then operated continuously. Equipment sized to meet the peak building power requirement is not likely to be economical unless power can be sold back to the utility at attractive rates.

Fuel Source. See discussion for office buildings.

Codes and Standards Compliance. See discussion for office buildings

Emissions. See discussion for office buildings.

Maintenance. Maintenance is likely to be performed through service contracts.

Chapter 5

CONCLUSIONS

Fuel cell systems promise to be an attractive on-site generation technology for incorporation in building cogeneration systems. Fuel cells offer high efficiency, minimal environmental impact, and applicability across a wide range of sizes. Within the next five years, products based on each of the four fuel cell technologies should be available. Price predictions for PEMFC, MCFC, and SOFC systems indicate that these systems will be available at prices that make them attractive in specific applications. Residential scale applications are likely to be served by PEMFC and SOFC technology. Commercial and industrial scale systems will most likely use MCFC or SOFC technology except in small sizes where PEMFC technology will be competitive.

Initial markets for fuel cell cogeneration are likely to be in areas like the Northeast where electricity costs are high. Initial applications will include hotels, which have high thermal loads, facilities such as large office buildings and grocery stores that have backup power requirements, and hospitals that have both auxiliary power requirements and high thermal loads. As fuel cell system costs decline, other opportunities should become attractive.

The successful application of fuel cell cogeneration systems in buildings requires selection of an appropriate technology, selection of the most attractive operating strategy, coordination with local utility rates and regulations, compliance with building regulations, and evaluation of the economic merit of the application. Economic analyses of typical commercial and residential applications indicate that cogeneration systems employing fuel cells can be economically attractive if the initial costs can be reduced to the range of \$1000 to \$1500/kWe. Current cost projections indicate that this range is achievable with a possibility of much lower costs should PEMFC technology become competitive in transportation applications.

Building designers are a critical link in the adoption of fuel cell cogeneration technology. By understanding fuel cells and cogeneration systems, designers can identify appropriate applications that will yield economic and environmental benefits for their clients and for society.

REFERENCES

1. Larminie, J., and A. Dicks. 2000. *Fuel cell systems explained.* Chichester: John Wiley & Sons, LTD.
2. Little, Arthur D. 1994. *Multi-fuel reformers for fuel cells used in transportation—Phase I.* Final report for U.S. Department of Energy, Office of Transportation technologies, DE-AC02-92-CE50343. Cambridge, Mass.
3. Matsumoto, M., and K. Usami. 1992. PAFC commercialization and recent progress of technology in Mitsubishi Electric. In International Fuel Cell Conference, Tokyo, Japan. NEDO/MITI.
4. Blomen, L.J.M.J., and M.N. Mugerwa. 1993. *Fuel cell systems.* New York: Plenum Press.
5. Kordesch, K., and G. Simader. 1996. *Fuel cells and their applications.* New York: Vch.
6. Marr, C., and X.G. Li. 1999. Composition and performance modeling of catalyst layer in a proton exchange membrane fuel cell. *Journal of Power Sources* 77(1): 17-27.
7. Mosdale, R., G. Gebel, and M. Pineri. 1996. Water profile determination in a running proton exchange membrane fuel cell using small-angle neutron scattering. *Journal of Membrane Science* 118(2): 269-277.
8. Eikerling, M., Y.I. Kharkats, A.A. Kornyshev, and Y.M. Volfkovich. 1998. Phenomenological theory of electro-osmotic effect and water management in polymer electrolyte proton-conducting membranes. *Journal of Electrochemical Society* 145(8): 2684-2699.
9. Verbrugge, M.W., and R.F. Hill. 1990. Ion and solvent transport in ion-exchange membranes: I. A macrohomogenous mathematical model. *Journal of Electrochemical Society* 137(3): 886-892.

10. Fuller, T.F., and J. Newman. 1993. Water and thermal management in solid-polymer-electrolyte fuel cells. *Journal of the Electrochemical Society* 140(5): 1218-1225.

11. Broka, K., and P. Ekdunge. 1997. Modelling the PEM fuel cell cathode. *Journal of Applied Electrochemistry* 27(3): 281-289.

12. Springer, T.E., T.A. Zawodzinski, and S. Gottesdfeld. 1991. Polymer electrolyte fuel cell model. *Journal of Electrochemical Society* 138(8): 2334-2342.

13. Lee, S.J., S. Mukerjee, J. McBreen, Y.W. Rho, Y.T. Kho, and T.H. Lee. 1998. Effects of Nafion impregnation on performances of PEMFC electrodes. *Electrochimica Acta* 43(24): 3693-3701.

14. Yi, J.S., and T.V. Nguyen. 1998. An along-the-channel model for proton exchange membrane fuel cells. *Journal of the Electrochemical Society* 145(4): 1149-1159.

15. Singh, D., D.M. Lu, and N. Djilali. 1999. A two-dimensional analysis of mass transport in proton exchange membrane fuel cells. *International Journal of Engineering Science* 37(4): 431-452.

16. Kim, J., S.M. Lee, S. Srinivasan, and C.E. Chamberlin. 1995. Modeling of proton-exchange membrane fuel-cell performance with an empirical-equation. *Journal of the Electrochemical Society* 142(8): 2670-2674.

17. Amplett, J.C., R.M. Baumert, R.F. Mann, B.A. Peppley, P.R. Rabërge, and T.J. Harris. 1995. Performance modeling of the Ballard Mark IV solid polymer electrolyte fuel cell: I. Mechanical model development. *Journal of Electrochemical Society* 142(1): 1-15.

18. Amphlett, J.C., R.M. Baumert, R.F. Mann, B.A. Peppley, P.R. Roberge, and A. Rodrigues. 1994. Parametric modeling of the performance of a 5-Kw proton-exchange membrane fuel-cell stack. *Journal of Power Sources* 49(1-3): 349-356.

19. Gurau, V., H. Liu, and S. Kakac. 1998. Mathematical model for proton exchange membrane fuel cells. In *Proceedings of ASME Advanced Energy Systems Division*, Vol. 38, pp. 205-214.

20. Bernardi, D.M., and M.W. Verbrugge. 1992. A mathematical model of the solid-polymer-electrolyte fuel cell. *Journal of Electrochemical Society* 139(9): 2477-2490.

21. Sishtla, C., G. Koncar, R. Platon, S. Gamburzev, A.J. Appleby, and O.A. Velev. 1998. Performance and endurance of a PEMFC oper-

ated with synthetic reformate fuel feed. *Journal of Power Sources* 71(1-2): 249-255.

22. Hirschenhofer, J.H., D.B. Stauffer, R.R. Engleman, and M.G. Klett. 1998. *Fuel Cell Handbook,* 4th Edition. Morgantown, W.V.: US-DOE, Office of Fossil Energy, FETC.

23. Zawodzinski, T.A., T.E. Springer, F. Uribe, and S. Gottesfeld. 1993. Characterization of polymer electrolytes for fuel-cell applications. *Solid State Ionics* 60(1-3): 199-211.

24. Prater, K.B. 1990. The renaissance of the solid polymer fuel cell. *Journal of Power Sources* 29(1-2): 239-250.

25. Watanabe, M., H. Uchida, Y. Seki, M. Emori, and P. Stonehart. 1996. Self-humidifying polymer electrolyte membranes for fuel cells. *Journal of the Electrochemical Society* 143(12): 3847-3852.

26. Ralph, T.R., G.A. Hards, J.E. Keating, S.A. Campbell, D.P. Wilkinson, M. Davis, J. StPierre, and M.C. Johnson. 1997. Low cost electrodes for proton exchange membrane fuel cells—Performance in single cells and Ballard stacks. *Journal of the Electrochemical Society* 144(11): 3845-3857.

27. Srinivasan, S., O.A. Velev, A. Parthasarathy, D.J. Manko, and A.J. Appleby. 1991. High-energy efficiency and high-power density proton-exchange membrane fuel-cells—Electrode-kinetics and mass-transport. *Journal of Power Sources* 36(3): 299-320.

28. Donitz, W. 1998. Fuel cells for mobile applications, status, requirements and future application potential. *International Journal of Hydrogen Energy* 23(7): 611-615.

29. Denis, M.C., G. Lalande, D. Guay, J.P. Dodelet, and R. Schulz. 1999. High energy ball-milled Pt and Pt-Ru catalysts for polymer electrolyte fuel cells and their tolerance to CO. *Journal of Applied Electrochemistry* 29(8): 951-960.

30. Oetjen, H.F., V.M. Schmidt, U. Stimming, and F. Trila. 1996. Performance data of a proton exchange membrane fuel cell using H-2/CO as fuel gas. *Journal of the Electrochemical Society* 143(12): 3838-3842.

31. Swette, L.L., A.B. Laconti, and S.A. McCatty. 1994. Proton-exchange membrane regenerative fuel-cells. *Journal of Power Sources* 47(3): 343-351.

32. Stimming, U., and K.A. Friedrich. 1998. *Brennstoffzelle mit pulsfoermig veraendertem Anodenpotential.* Germany patent DE 197 10 819 C1.

33. Murphy, O.J., A. Cisar, and E. Clarke. 1998. Low-cost light weight high power density PEM fuel cell stack. *Electrochimica Acta* 43(24): 3829-3840.

34. Brown, D.R. 1998. *PEM Fuel Cells for Commercial Buildings.* Pacific Northwest National Laboratory. Prepared for U.S. DOE, DE-AC06-76RLO 1830.

35. Ballard Power Systems Inc. 1999. Ballard Power Systems home page. Available online from <http://www.ballard.com/>.

36. Prater, K.B. 1992. Solid polymer fuel-cell developments at Ballard. *Journal of Power Sources* 37(1-2): 181-188.

37. Weiner, S.A. 1998. Fuel cell stationary power business development. *Journal of Power Sources* 71(1-2): 61-64.

38. Amphlett, J.C., R.F. Mann, B.A. Peppley, P.R. Roberge, A. Rodrigues, and J.P. Salvador. 1998. Simulation of a 250 kW diesel fuel processor PEM fuel cell system. *Journal of Power Sources* 71(1-2): 179-184.

39. PlugPower Inc. 1999. PlugPower home page. Cited 1999. Available online from <http://www.plugpower.com/>.

40. USCAR. 1999. USCAR home page. Cited 1999. Available online from <http://www.uscar.org/>.

41. Prater, K.B. 1994. Polymer electrolyte fuel-cells—A review of recent developments. *Journal of Power Sources* 51(1-2): 129-144.

42. Prater, K.B. 1996. Solid polymer fuel cells for transport and stationary applications. *Journal of Power Sources* 61(1-2): 105-109.

43. Barbir, F., and T. Gomez. 1997. Efficiency and economics of proton exchange membrane (PEM) fuel cells. *International Journal of Hydrogen Energy* 22(10-11): 1027-1037.

44. Teagan, W.P., J. Bentley, and B. Barnett. 1998. Cost reductions of fuel cells for transport applications: fuel processing options. *Journal of Power Sources* 71(1-2): 80-85.

45. Aragane, J., H. Urushibata, and T. Murahashi. 1996. Effect of operational potential on performance decay rate in a phosphoric acid fuel cell. *Journal of Applied Electrochemistry* 26(2): 147-152.

46. Appleby, A.J., and F.R. Foulkes. 1989. *Fuel cell handbook*. New York: Van Nostrand Reinhold.

47. Song, R.H., D.R. Shin, and C.S. Kim. 1998. New method of electrode fabrication for phosphoric acid fuel cell. *International Journal of Hydrogen Energy* 23(11): 1049-1053.

48. ONSI Corporation. 2000. ONSI Corporation home page. Cited 2000. Available online from <http://www.onsicorp.com/>.

49. Gibbs, C.E., and M.C.F. Steel. 1992. European opportunities for fuel cell commercialization. *Journal of Power Sources,* Vol. 37, pp. 35-43.

50. Matsumoto, K., and K. Kasahara. 1998. Long-term commitment of Japanese gas utilities to PAFCs and SOFCs. *Journal of Power Sources* 71(1-2): 51-57.

51. Nymoen, H. 1994. PAFC Demonstration plants in Europe—First results. *Journal of Power Sources* 49(1-3): 63-76.

52. Gibbs, C.E., and M.C.F. Steel. 1992. European opportunities for fuel-cell commercialization. *Journal of Power Sources* 37(1-2): 35-43.

53. Hojo, N., M. Okuda, and M. Nakamura. 1996. Phosphoric acid fuel cells in Japan. *Journal of Power Sources* 61(1-2): 73-77.

54. Nakajima, N., H. Miyoshi, K. Nishizaki, T. Kitai, H. Uchida, and M. Watanabe. 1997. Evaluation of the performance degradation of phosphoric acid fuel cells, dealloying process of Pt-Co-Ni electrocatalysts. *Denki Kagaku* 65(10): 876-877.

55. Appleby, A.J. 1996. Issues in fuel cell commercialization. *Journal of Power Sources,* Vol. 69, pp. 153-176.

56. Yasue, H., H. Kato, and K. Takasu. 1998. Development of a 1000 kW class MCFC pilot plant in Japan. *Journal of Power Sources* 71(1-2): 89-94.

57. Baker, B., S. Gionfriddo, A. Leonida, H. Maru, and P. Patel. 1984. Internal reforming natural gas fueled carbonate fuel cell stack. Energy Research Corporation. Final report for Gas Research Institute, 5081-244-0545, March.

58. Benjamin, T.G., E.H. Caramar, and L.G. Marianowski. 1980. *Handbook of fuel cell performance*. Institute of Gas Technology.

59. Lu, S.H., and J.R. Selman. 1984. In: Symposium on molten carbonate fuel cell technology, J.R. Selman and T.D. Claar, eds. *Proceedings of the Electrochemical Society*, Vol. 84:13. Pennington, N.J.: Electrochemical Society.

60. Hioshino, K., and T. Kohno. 1992. Development of copper base anodes for molten carbonate fuel cells. In *International Fuel Cell Conference Proceedings*, Tokyo, Japan, pp. 169-172.

61. Shimada, T., T. Fukui, K. Hato, S. Terada, E. Hoshino, K. Ohe, M. Yamamoto, M. Hori, I. Nagashima, and A. Miki. 1996. New materials for molten carbonate fuel cell. *Denki Kagaku* 64(6): 533-541.

62. Joon, K. 1996. Critical issues and future prospects for molten carbonate fuel cells. *Journal of Power Sources* 61(1-2): 129-133.

63. Freni, S., S. Cavallaro, M. Aquino, D. Ravida, and N. Giordano. 1994. Lifetime-limiting factors for a molten-carbonate fuel-cell. *International Journal of Hydrogen Energy* 19(4): 337-341.

64. Bosio, B., P. Costamagna, F. Parodi, and B. Passalacqua. 1998. Industrial experience on the development of the molten carbonate fuel cell technology. *Journal of Power Sources* 74(2): 175-187.

65. Tanimoto, K., M. Yanagida, T. Kojima, Y. Tamiya, H. Matsumoto, and Y. Miyazaki. 1998. Long-term operation of small-sized single molten carbonate fuel cells. *Journal of Power Sources* 72(1): 77-82.

66. Freni, S., M. Aquino, and E. Passalacqua. 1994. Molten-carbonate fuel-cell with indirect internal reforming. *Journal of Power Sources* 52(1): 41-47.

67. Passalacqua, E., S. Freni, F. Barone, and A. Patti. 1996. Porous ceramic membranes for direct internal reforming molten carbonate fuel cells. *Materials Letters* 29(1-3): 177-183.

68. Hirschenhofer, J.H. 1997. Fuel cell status: 1996. *IEEE Aerospace and Electronic Systems Magazine* 12(3): 23-28.

69. Watanabe, Y., M. Matsumoto, and K. Takasu. 1996. The market for utility-scale fuel cell plants. *Journal of Power Sources* 61(1-2): 53-59.

70. Sasaki, H., S. Otoshi, M. Suzuki, T. Sogi, A. Kajimura, N. Sugiura, and M. Ippommatsu. 1994. Fabrication of high-power density tubular type solid oxide fuel-cells. *Solid State Ionics*, Vol. 72, pp. 253-256.

71. Bevc, F. 1997. Advances in solid oxide fuel cells and integrated power plants. In *Proceedings of the Institution of Mechanical Engineers, Part A—Journal of Power and Energy*, Vol. 211, pp. 359-366.

72. Casanova, A. 1998. A consortium approach to commercialized Westinghouse solid oxide fuel cell technology. *Journal of Power Sources* 71(1-2): 65-70.

73. Achenbach, E. 1995. Response of a solid oxide fuel cell to load change. *Journal of Power Sources* 57(1-2): 105-109.

74. Kilbride, I.P. 1996. Preparation and properties of small diameter tubular solid oxide fuel cells for rapid start-up. *Journal of Power Sources* 61(1-2): 167-171.

75. Bessette, N.F., and W.J. Wepfer. 1996. Prediction of on-design and off-design performance for a solid oxide fuel cell power module. *Energy Conversion and Management* 37(3): 281-293.

76. Huijsmans, J.P.P., F.P.F. van Berkel, and G.M. Christie. 1998. Intermediate temperature SOFC—A promise for the 21st century. *Journal of Power Sources*, Vol. 71, pp. 107-110.

77. Dollard, W.J. 1992. Solid oxide fuel-cell developments at Westinghouse. *Journal of Power Sources* 37(1-2): 133-139.

78. Eguchi, K., Y. Kunisa, M. Kayano, K. Sekizawa, S. Yano, and H. Arai. 1996. Power generation characteristics of solid oxide fuel cell with internal steam reforming of methane. *Denki Kagaku* 64 (6): 596-601.

79. Meusinger, J., E. Riensche, and U. Stimming. 1998. Reforming of natural gas in solid oxide fuel cell systems. *Journal of Power Sources* 71(1-2): 315-320.

80. Kawada, T., and H. Yokokawa. 1997. Materials and characterization of solid oxide fuel cell. In *Electrical Properties of Oxide Materials*. Clausthal Zellerfe: Trans Tech Publications, pp. 187-248.

81. Bakker, W., and R. Goldstein. 1996. Development of low temperature solid oxide fuel cells. In *1996 Fuel Cell Seminar*, pp. 48-50.

82. Sammells, A.F. 1992. Perovskite solid electrolytes for SOFC, in *Proceedings of the Fourth Annual Fuel Cells Contractors Review Meeting*, July. U.S. DOE/METC.

83. Feng, M., J.B. Goodenough, K. Huang, and C. Milliken. 1996. Fuel cells with doped lanthanum gallate electrolyte. *Journal of Power Sources*, Vol. 63, pp. 47-51.

84. Singhal, S.C. 1997. Recent progress in tubular solid oxide fuel cell technology. In *Proceedings of the Fifth International Symposium on Solid Oxide Fuel Cells (SOFC-V)*. Pennington, N.J.: The Electrochemical Society, Inc.

85. Choy, K., W. Bai, S. Clarojrochkul, and B.C.H. Steele. 1998. The development of intermediate-temperature solid oxide fuel cells for the next millennium. *Journal of Power Sources* 71(1-2): 361-369.

86. Drenckhahn, W. 1996. Solid-oxide fuel cells and CHP—An environmental solution in the making. *Power Engineering Journal* 10(2): 67-72.

87. Song, R.H., T. Horita, N. Sakai, T. Kawada, H. Yokokawa, and M. Dokiya. 1996. Fabrication of planar solid oxide fuel cell by composite plate process. *Denki Kagaku* 64(6): 614-619.

88. Elangovan, S., J. Hartvigsen, A. Khandkar, R.M. Privette, K.E. Kneidel, M.A. Perna, and D.R. Rowley. 1998. Planar solid oxide fuel cell integrated system technology development. *Journal of Power Sources* 75(1): 185-185.

89. Winkler, W., and J. Kruger. 1998. Design and manufacturing of a tubular solid oxide fuel cell combustion system. *Journal of Power Sources* 71(1-2): 244-248.

90. Bessette, N.F., and R.A. George. 1996. Performance and reliability of Westinghouse's air electrode supported solid oxide fuel cell at atmospheric and elevated pressures. *Denki Kagaku* 64(6): 602-608.

91. Brown, J.T. 1988. Solid oxide fuel cell technology. *IEEE Transactions on Energy Conversion* 3(2): 193-198.

92. Federal Energy Technology Center. 1998. *Solid oxide fuel cell project: Generating tomorrow's electricity cleanly.* U.S. DOE Office of Fossil Energy, FETC, Morgantown, W.V. Project facts, PS 026 1098M.

93. George, R.A., and N.F. Bessette. 1998. Reducing the manufacturing cost of tubular SOFC technology. *Journal of Power Sources* 71(1-2): 131-137.

94. Itoh, H., M. Mori, N. Mori, and T. Abe. 1994. Production cost estimation of solid oxide fuel-cells. *Journal of Power Sources* 49(1-3): 315-332.

95. Ippommatsu, M., H. Sasaki, and S. Otoshi. 1996. Evaluation of the cost performance of the SOFC cell in the market. *International Journal of Hydrogen Energy* 21(2): 129-135.

96. DeBlasio, R. 2000. Status update: IEEE SCC21 activities. In *Fuel Cell Summit IV*, Washington, D.C., May 10-11.

97. Simpkins, E. 2000. Fuelcell energy—Status of MCFC development. In *Fuel Cell Summit IV*, Washington, D.C., May 10-11.

98. Shibata, K., and K. Watanabe. 1994. Philosophies and experiences of PAFC field trials. *Journal of Power Sources* 49(1-3): 77-102.

99. U.S. Department of Defense. 2000. U.S. DOD Fuel Cell Demonstration Program home page. Cited Nov 19, 2000. Available online from http://www.dodfuelcell.com.

100. Teagan, W.P., and D.J. Friedman. 1994. Fuel cells for building applications: Definition of performance/cost requirements. In *Intersociety Energy Conversion Engineering Conference*, Vol. 3. Monterey, Calif., August, IEEE, pp. 1251-1257.

101. Aspen Systems Corporation. 2000. *Combined heat & power: A federal manger's resource guide*. Aspen Systems Corporation, Rockville, Md. Final report for U.S. DOE Federal Energy Management Program, March.

102. U.S. Department of Energy—Energy Information Administration. 2000. *1997 Residential energy consumption survey*. Cited Jan 2000. Available online from http://www.eia.doe.gov/emeu/recs.

103. U.S. Department of Energy—Energy Information Administration. 2000. *1995 Commercial buildings energy consumption survey*. Cited Jan 2000. Available online from http://www.eia.doe.gov/emeu/cbecs.

104. U.S. Fuel Cell Council. 2000. *Fuel Cell Glossary*, 2nd Ed. Cited Nov 2000. Available online from http://www.usfcc.com/Glossary2.pdf.

105. Orlando, J.A. 1996. *Cogeneration design guide*. Atlanta: American Society of Heating, Refrigerating and Air-Conditioning Engineers, Inc.

Appendix A

ADDITIONAL RESOURCES

FUEL CELL SYSTEM REFERENCES

Books

Fuel Cell Handbook by A.J. Appleby and F.R. Foulkes, Van Nostrand Reinhold, New York, 1989.

Fuel Cell Systems by L.J. Blomen and M.N. Mugerwa, Plenum Press, New York, 1993.

Fuel Cell Handbook, 5th ed. by EG&G Services, Parsons, Inc., Science Applications International Corporation, US-DOE, Office of Fossil Energy, National Energy Technology Laboratory, Morgantown, W.V., 2000.

Fuel Cells and Their Applications, K. Kordesch and G. Simader (VCH, New York, 1996).

Websites

Fuel Cells 2000	<www.fuelcells.org>
Electric Power Research Institute	<www.epri.com>
Gas Research Institute	<www.gri.org>
U.S. CAR	<www.uscar.org>
DOE R&D Project Summaries	<www.doe.gov/rnd/dbhome>
DOE CHP Challenge	<www.eren.doe.gov/der/chp/>
U.S. DOD Fuel Cell Demonstration Program	<www.dodfuelcell.com>
Argonne National Labs	<www.anl.gov>
Los Alamos National Labs	<www.lanl.gov>
National Energy Technology Laboratories	<www.fetc.doe.gov>

Sandia National Labs	<www.sandia.gov>
National Fuel Cell Research Center	<www.nfcrc.uci.edu>

MANUFACTURERS

An updated list of fuel cell manufacturers along with links to their websites is maintained at the Fuel Cell 2000 web site: <www.fuelcells.org>.

Appendix B

GLOSSARY

anode[1] — The negative electrode at which oxidation occurs.

balance-of-plant — Auxiliary components and subsystems such as fuel processing, thermal management, water management, and power conditioning that are required to operate a fuel cell system.

bipolar plates[1] — Conductive plate in a fuel cell stack, which acts as an anode for one cell and a cathode for the adjacent cell. The plate usually incorporates flow channels for the fluid feeds and may also contain conduits for heat transfer.

capacity factor[2] — The ratio of the actual annual plant electricity output to the rated plant output (fraction or percent).

catalyst — A chemical substance (often a noble metal) that increases the rate of a chemical reaction without participating in the reaction. The catalyst lowers the activation energy for the reaction, thus allowing it to take place faster or at a lower temperature.

1. Definition is from the *Fuel Cell Gossary* published by U.S. Fuel Cell Council [104].
2. Definition is from the *Cogeneration Design Guide* [105].

catalyst loading	The mass of catalyst per unit area of the fuel cell electrode (mg/cm^2).
cathode	The positive electrode at which reduction occurs.
cogeneration	The simultaneous production of electricity or mechanical work and useful thermal energy from a single energy source.
cogeneration efficiency	The ratio of the total useful energy provided by the system (including both work and thermal energy) to the energy input to the system.
compressor	Rotary (turbocompressor) or reciprocating device for using power to increase the pressure of a gas stream.
current collector	Microporous conductive materials located between the bipolar plate and the electrode. The current collector helps to distribute current to (or from) the bipolar plate to the surface of the electrode.
current density	Current flow per unit area of the fuel cell electrolyte (mA/cm^2).
desulfurizer	A reactor (typically filled with activated carbon or zinc oxide) that is used to remove sulfur from a fuel gas stream.
diffusion	Species transport that is due to a gradient in chemical potential or concentration (gmol/s).
direct internal reforming	Conversion of a hydrocarbon fuel by steam reforming and shift conversion reactions that occur within the anode chamber of the fuel cell stack.
direct methanol fuel cell[1] (DMFC)	A type of fuel cell in which the fuel is methanol (CH_3OH) in gaseous or liquid form. The methanol is oxidized directly at the anode with no reformation to hydrogen. The electrolyte is typically a polymer electrolyte.

distributed generation — Generation of electricity (usually at a small scale) near the point of use. Common distributed generation systems include internal combustion engines, gas turbines, photovoltaics, wind energy, and fuel cells.

efficiency — The ratio of the desired energy output to the energy input. See *electrical conversion efficiency*, *fuel processor efficiency*, and *cogeneration efficiency*.

electrical conversion efficiency — The ratio of the net electrical energy provided by the fuel cell system to the energy released by the complete reaction of the fuel entering the fuel cell system.

electrode — An electrically conductive component where either the oxidation or reduction half-reactions take place.

electro-osmotic drag[1] — The flux of a polar species such as water due to its attraction to a proton (H^+) that is transported from the anode to the cathode.

electrolyte — An ionic conductor that is resistant to the flow of electrons and the flow of reactants.

expander — Rotary (turboexpander) or reciprocating device for expanding a gas stream to extract power while reducing the gas pressure.

external reforming — The conversion of a hydrocarbon fuel to fuel stream rich in H_2 (or H_2 and CO for MCFC and SOFC cells) that occurs outside of the fuel cell stack.

Faraday constant — The amount of charge contained in a mole of electrons:

$F = 96{,}487$ coulombs/gmol-electrons

fuel cell[2]	An electrochemical device that can continuously convert the chemical energy of a fuel and an oxidant to electrical energy.
fuel processor efficiency	The ratio of the heating value of the fuel entering the fuel processor to the heating value of the fuel leaving the fuel processor.
fuel utilization	The ratio of fuel reacted in the fuel cell to the fuel supplied for reaction $[(fuel_{in} - fuel_{out}) / fuel_{in}]$.
liquefied petroleum gases	Hydrocarbon fuel that is composed primarily of propane, propylene, butanes, and butylenes.
natural gas	Hydrocarbon fuel that is composed primarily of methane (CH_4).
grid connected	A mode of operation for a distributed generation system in which the distributed generation system is interconnected with the utility power system. Grid-connected systems may supply some or all of the electricity required by the user and may supply energy back to the utility grid.
grid isolated	A mode of operation for a distributed generation system in which the distributed generation system is not interconnected with the utility power system. Grid-isolated systems must supply all of the electricity required by the user and cannot supply energy back to the utility grid.
gross fuel cell power	The DC power provided by the fuel cell stack (kWe). Gross power does not reflect any power conditioning losses or parasitic power requirements by system components such as pumps and fans.

heat rate — A measure of the performance of a power plant that is defined as the ratio of the heating value of the fuel input in Btu's to the energy output in kWh (Btu/kWh). The heat rate is based on either the LHV or the HHV of the fuel.

higher heating value (HHV) — The energy released during the complete oxidation of a fuel where the products include water in the liquid form (kJ/kmol or Btu/lbmol).

indirect internal reforming — Conversion of a hydrocarbon fuel by steam reforming reactions that occur adjacent to the anode chamber of the fuel cell stack so that heat transfer from the cell reaction can be used to drive the endothermic reforming reaction.

load following — A mode of operation in which the power (electric load following) or thermal output (thermal load following) of the fuel cell varies to match the power or thermal needs of the user.

lower heating value (LHV) — The energy released during the complete oxidation of a fuel where the products include water in the vapor form (kJ/kmol or $Btu/lb_m mol$).

membrane electrode assembly (MEA) — For a proton exchange membrane fuel cell, the MEA is the combined structure consisting of the polymer membrane, the catalyst layers, the electrodes, and the gas diffusion layers.

molten carbonate fuel cell[1] (MCFC) — A type of fuel cell consisting of a molten electrolyte of Li_2CO_3/Na_2CO_3 in which the species CO_3^{2-} is transported from the cathode to the anode. Operating temperatures are typically near 650°C (1,200°F).

naphtha[1] An artificially produced petroleum or coal tar fraction with a volatility between gasoline and kerosene.

net power (a) The power available to the user from the fuel cell system after subtracting power conditioning losses and power requirements for auxiliary equipment such as fans and pumps (kWe).

(b) The difference between the power supplied *by the grid* to the site and the power supplied *to the grid* by the site.

open circuit voltage The voltage across the fuel cell when there is no current flow (V).

partial oxidation Fuel reforming process in which a portion of the hydrocarbon fuel is combusted to provide heat to drive the endothermic steam reforming reaction.

phosphoric acid fuel cell[1] **(PAFC)** A type of fuel cell in which the electrolyte consists of concentrated phosphoric acid (H_3PO_4) and protons (H^+) are transported from the anode to the cathode. The operating temperature range is generally 160-220°C (320-430°F).

polarization curve A plot of fuel cell voltage versus current density that describes the fuel cell performance for a given set of operating conditions.

polymer electrolyte fuel cell (PEFC) See *proton exchange membrane fuel cell.*

power conditioning The subsystem that converts DC power from the fuel cell stack to AC power with the proper voltage and power factor for connection to the user and/or the utility grid.

power density	(a) The power produced per unit area of the fuel cell electrolyte (W/cm^2). (b) The power produced per unit volume of a fuel cell stack or system (kWe/liter).
power factor	The ratio of the real power in watts to the apparent power in volt-amperes for an AC power circuit.
preferential oxidation	A catalyzed reaction in which one species is oxidized more rapidly than another. For example, in fuel cell systems, a preferential oxidation system is often used to oxidize CO in the fuel stream to CO_2 while H_2 remains un-oxidized. Also called "selective oxidation."
proton exchange membrane	The polymer membrane that serves as the electrolyte in a proton exchange membrane fuel cell.
proton exchange membrane fuel cell (PEMFC)	A type of fuel cell in which the electrolyte consists of a polymer membrane that conducts protons (H^+) from the anode to the cathode. The operating temperature range is generally 50-90°C (120-190°F). Also called a *polymer electrolyte fuel cell* (PEFC) and a *solid polymer fuel cell* (SPFC).
reformer	A reactor, usually filled with a catalyst, in which a hydrocarbon fuel is reacted with water, heat, and possibly air to convert the hydrocarbon fuel to a fuel stream that is rich in H_2 and CO.
reformate gas	The fuel stream that leaves a reformer.
separator plate[1]	A solid piece of electrically conductive material (usually a metal or graphite) that is inserted between cells in a stack.

shift conversion	The exothermic reaction of CO and water to produce hydrogen and carbon dioxide, usually occurring over a catalyst.
system efficiency	The ratio of the net power output of the fuel cell system to the net power output of the fuel cell stack.
solid oxide fuel cell[1] **(SOFC)**	A type of fuel cell in which the electrolyte is a solid, nonporous metal oxide, typically ZrO_2 doped with Y_2O_3 and O^{2-}, that is transported from the cathode to the anode. The temperature of operation is typically 800-1,000°C (1,500-1,800°F).
solid polymer fuel cell (SPFC)	See *proton exchange membrane fuel cell.*
stack life[1]	The cumulative period of time that a fuel cell stack may operate before its output deteriorates below a useful minimum value.
steam reforming	An endothermic fuel reforming process in which a mixture of a hydrocarbon fuel and steam is heated and reacted over a catalyst to produce H_2 and CO. Steam reforming is most applicable to lighter hydrocarbons such as methane (CH_4).
thermal efficiency	Ratio of the power output of a system to the rate of energy input.
thermal management	Within a fuel cell system, thermal management refers to the transfer of energy from process streams that are to be cooled to process streams that are to be heated.